Revolution in Lamps

A Chronicle of 50 Years of Progress

Second Edition

Revolution in Lamps

A Chronicle of 50 Years of Progress

Second Edition

Raymond Kane and Heinz Sell, Editors

River Publishers

Routledge
Taylor & Francis Group
LONDON AND NEW YORK

Published 2020 by River Publishers
River Publishers
Alsbjergvej 10, 9260 Gistrup, Denmark
www.riverpublishers.com

Distributed exclusively by Routledge
4 Park Square, Milton Park, Abingdon, Oxon OX14 4RN
605 Third Avenue, New York, NY 10017, USA

Library of Congress Cataloging-in-Publication Data

Kane, Raymond.
Revolution in lamps : a chronicle of 50 years of progress/
Raymond Kane and Heinz Sell, editors.--2nd ed.
p. cm.
Includes bibliographical references and index.
ISBN 978-8-7702-2222-8 (electronic)
1. Incandescent lamps--History. I. Kane, Raymond. II. Sell, Heinz.

TK4351 .R48 2001
621.32′6′09--dc21

00-066250

Revolution in lamps; a chronicle of 50 years of progress by Raymond Kane and Heinz Sell
First published by Fairmont Press in 2001.

Routledge is an imprint of the Taylor & Francis Group, an informa business

0-88173-378-4 (The Fairmont Press, Inc.)
978-0-8247-0917-4 (print)
978-8-7702-2222-8 (online)
978-1-0031-5098-5 (ebook master)

Contents

Chapter

Preface to the Second Edition

This book is written for designers, engineers, architects, students and others, both professional and lay, interested in the development of electric light sources from World War II to the near present.

While other books and publications have covered new lamp technologies to some extent, none has given this period and its many advancements a comprehensive accounting. It is the intent of *Revolution in Lamps* to accomplish this via the contributions of editors Ray Kane and Heinz Sell and their colleagues, who all shared in exciting lamp and lamp materials development over more than 30 years. As the authors predominantly worked at Westinghouse/Philips and OSRAM, the material is provided from the point of view of their experience at these companies, although its intent is to be non-biased and comprehensive. *Revolution in Lamps* also serves as an invaluable historical document adding to the industry's body of knowledge and contributing to its growing sense of history and culture.

It is important to view the historical context of the various lamps because the new developments of today have their origins in the lamps of the past. Life cycles can also be observed. Over the years, significant new developments have been followed by periods of maturation of the new source type, and then an evolution of the basic design into a broader, deeper product line with increased application opportunities. For example, from limited initial application, tungsten halogen sources today dominate incandescent retail display lighting. Fluorescent lamp technology has allowed an increasing choice of compact lamp configurations and because of the use of rare earth phosphors, high color rendering sources. Some types mature and lose market share to improved lamps within the same family. Mercury vapor lamps, once popular for roadway and industrial lighting designs, are seldom specified. Metal halide and high pressure sodium lamps are the sources of choice for these applications. The current metal halide family of lamps has overcome the early problems of color deficiency (reds) and tendency to color shift over life. Good color rendering, color stability and smaller envelope sizes enable metal halide lamps to be used in recessed and track lighting application in commercial spaces.

Market trends and issues influence the design, development and availability of source types. Energy and environmental concerns have changed consumer attitudes and buying habits. The consumer today is knowledgeable about energy management, energy costs, product efficiency, and conservation of the environment. Regulation has limited manufacturing and importing of certain types of lamps, placed efficacy requirements on lamps that are brought to market, and exercised control over the way we dispose of spent lamps.

The reader will learn not only about various light sources, their function and development in the post-World War II era, but also the authors' thoughts about the future of each lamp type. Among the products that have shown potential for growth in the next few years are LEDs, first introduced in the 1960s, with a "next generation" improvement in the 1990s, electrodeless lamps, which appeared in the 1980s, and fiber-optic designs. Continued improvement in circuitry and ballasts will also enable increased lamp efficacy. Lamp types for dedicated applications will continue to be developed. They satisfy needs for special effects, for non-visual purposes, for health, and for high performance applications such as projection TV.

The reader should recognize the importance of the light source among all the components of a lighting system. It is the heart of the system. It is selected before the luminaires, ballasts or control devices, and determines the quality and performance of the system and all the other design and equipment choices that follow.

The authors are pleased to convey their experiences and knowledge to the reader, and welcome any comments on this book.

Acknowledgments

Any book of the scope attempted here must fall short in giving recognition to the efforts and achievements of all of the engineers who, as members of the lamp industry, have contributed in a myriad of ways to the "revolution in lamps." While the authors believe that they have acknowledged the contributions of many of their colleagues in the references listed in each chapter, they recognize that a broader endorsement is warranted.

The revolution in lamps would not have been possible without the ingenious equipment designs conceived by the mechanical, electrical and electronic engineers who have carried lamp manufacturing to a remarkable state of refinement, complexity and speed. It was they who designed the very intricate mechanical devices that made possible the mass production of the new light sources. While in the early days, following the introduction of the tungsten coil and coiled-coil filaments, incandescent lamps were automatically assembled on indexing machines at a rate of a few hundred per hour. Today, such lamps are produced at the phenomenal rate of more than 9,000 per hour and with unparalleled quality. In addition, it seems incomprehensible that the filaments for these lamps are wound on molybdenum mandrels at the fantastic rate of more than 30,000 windings per minute.

As remarkable as these accomplishments are, the extraordinary designs of incandescent lamp manufacturing equipment have been eclipsed by the sophisticated automated machines on which the more complex fluorescent and HID lamps are produced nowadays. Some of these lamps are assembled at a rate exceeding 6,000 lamps per hour. And as to miniature HID lamps with quartz envelopes that must be processed at a melting temperature of more than 2,000°C, the production rates are commonly greater than 1,000 per hour. Some of the outstanding engineers who were involved in the design of this amazing equipment may have found their names on patents, but few publications sang their praise.

On the periphery of lamp development, vendors of special materials deserve to be recognized. To name just a few: Corning Glass Works with its prolific offering of glass tubing and glass bulbs produced on its ribbon machines at a rate of 60,000 per hour, General Electric with its high-purity quartz tubing and translucent sintered alumina, GEC England with the invention of the halophosphate phosphor, Anderson Physics Laboratory with its high-purity metal halide powders and pellets, Philips with the invention of the green (CAT) and blue (BAM) triphosphor components, and Bausch and Lomb with its dichroic and cold mirror reflectors.

Of course, the revolution in lamps was much influenced by the people in marketing. They often identified applications and encouraged developments which would not otherwise have been undertaken.

We consider it a privilege to extend this acknowledgment to all who have played a role in the revolution in lamps.

—*The Authors*

Contributors

EDITORS

Raymond M. Kane

Raymond Kane, PE received a BS degree in electrical engineering from the New Jersey Institute of Technology and an MS degree in mathematics from the Stevens Institute of Technology. Until his retirement in 1987, he has worked as a Fellow Engineer for the Westinghouse Lamp Division and Philips Lighting Company, where he focused on the design, development and application of incandescent, tungsten halogen, fluorescent and HID lamps. He was active in initiating Westinghouse's first production of T3 quartz, metal halide and high pressure sodium lamps. From 1968 to 1974, he served on the HID Lamp and Ballast Committees of ANSI. Most recently, Mr. Kane has worked as a lamp consultant to the government of South Korea.

Heinz G. Sell

Heinz Sell is a graduate of Munich University with a master's degree in physics awarded in 1951. Following graduation, he spent two years as a research physicist working on radiation dosimetry, one year as a project manager for a photometry and lamp development laboratory, and 34 years with Westinghouse Lamp Division/Philips Lighting Company. At Westinghouse/Philips, he worked initially as electrical engineer and senior research engineer; then, for 24 years, he served in key management positions for material and process development, including lamp grade tungsten wire and the development of tungsten halogen, metal halide and high pressure sodium lamps and corporate quality assurance. Mr. Sell has authored and co-authored 22 scientific and technical publications and holds six U.S. patents. Now retired, he is active as a consultant on lamp and material problems.

REVIEWERS

Rita M. Harrold, FIES

Rita Harrold is Director of Educational and Technical Development for the Illuminating Engineering Society of North America. Responsibilities include development of the IESNA educational program materials and technical documents.

Previously, she served as Manager, Sales Training for the Philips Lighting Company and Manager, Sales Training and Development for the Genlyte Group. She also worked for 15 years as Director of Lighting Design for Westinghouse's Commercial Division. At the same time, Ms. Harrold chaired numerous IESNA committees, including the Energy Management committee, for which she received an award from ASHRAE for contributions to Standard 90-1975, *Energy Conservation in New Buildings.*

Ms. Harrold, the 81st president of the IESNA (1985-86), was awarded Fellow status in 1981.

Thomas Madonia, PE

Tom Madonia is a lighting and training consultant; previously, he was a technical and training consultant for Philips Lighting Company, a lighting consultant for Syska & Hennessy, and a technical and marketing consultant for Magnalite Manufacturing Corp. Prior to his consulting experience, Mr. Madonia held a variety of marketing management positions for Westinghouse Electric/ Philips Lighting Company for 34 years. He is a professional engineer.

Mr. Madonia has served on a number of IESNA committees, including Finance (chairman), Maintenance (chairman) and Public Conveyances (chairman). He is a recipient of the IESNA's Distinguished Service Award. In addition, Mr. Madonia has written a number of articles on lighting topics, has taught lighting courses at Newark College of Engineering and New York University and for IES lighting programs, and has served on the NEMA Lamp Committee (chairman).

AUTHORS

Jacob Van Broekhoven

Jacob Van Broekhoven, PE devoted most of his 34 years as an engineer in the Westinghouse Lamp Division to the field of phosphor development. His efforts began when calcium halophosphate phosphors replaced the toxic zinc beryllium silicate, and continued to include the rare earth phosphors that are now finding extensive application. Beginning with the findings of the lamp division researchers, he successfully optimized many phosphors and developed manufacturing processes, including the preparation of manufacturing specifications and equipment selection. His experience covers most of the fluorescent and HID lamp phosphors, as well as some specialized cathode ray phosphors. He was responsible for the development and the final manufacturing of electroluminescent phosphors. Since retiring from Westinghouse, he has been involved in scandium chemistry and consulting. He earned his BS degree in chemical engineering from the New Jersey Institute of Technology in 1948 and his MS degree in metallurgy from the Stevens Institute of Technology in 1954.

Chester Dawson

Chester Dawson, formerly Manager of Incandescent Lamp Engineering and Director of R&D for the Westinghouse Lamp Division and Philips Lighting Company, is presently Director of Compact Disc Technology for Sony Music. Upon completion of his graduate studies at Cornell University, he began his career at Westinghouse in 1969, conducting basic research on bubble growth and migration in tungsten wire. This work led to studies of various filament failure mechanisms that were expanded to include development of an improved electrostatic lamp coating, better use of maintenance getters, and a VO_2 thermistor to limit the in-rush current when a lamp is switched on.

Alexander Dobrusskin, Dipl.Ing. Physicist

Mr. Dobrusskin is Vice President for Technology Services for factories and Total Quality Management for Osram in Germany. He has 29 years of research and development experience and was responsible for the development of HID lamps. He has been with Osram since 1960. Mr. Dobrusskin studied physics at the Technische Universitat Berlin.

Dr. Alfred Gahn

Dr. Gahn is presently the Senior Director of the Prematerial Development Department of Osram in Germany. He joined the company as a research engineer in its Metals Research Department in 1968 while completing his studies in chemistry, first at the University of Karlsruhe and starting in 1977 at Munich's Maximilian University. At Munich University, he was awarded a master of science degree in chemistry in 1982, followed by a Ph.D. in solid state chemistry in 1986. Dr. Gahn's research activities involved the chemistry and metallurgy of tungsten wire and electrodes for special lamp applications, as well as the utilization of phosphors, primarily in discharge lamps. He is the holder of several patents related to the production of tungsten wire.

Dr. Jürgen Heider

Dr. Heider, with Osram in Germany, is responsible for the Department of Expert Systems which is in search of optimized knowledge transfer. Joining Osram in 1978, he became manager of the development of metal halide lamps in 1993. Dr. Heider studied experimental physics and received his doctorate at the Ludwig Maximilans Universitat in Munich. During his study, he worked on electron beam-plasma interactions and the instabilities that are produced in these collisional plasma systems.

Dr.-Ing. Dieter Hofmann

Dr. Hofmann has been head of Osram's Central Lamp Development section in Munich since 1988. He has served Osram since 1967, making contributions (as Head of the Department for Fluorescent Lamp Development) to the improvement of fluorescent lamps and to the development of Osram compact fluorescent lamps. Dr. Hofmann studied physics at the Technical University of Berlin-Charlottenburg; his thesis was "Fractional Analysis of Tap Spectra in Phosphors."

Daniel A. Larson

Mr. Larson began his career at the General Electric Research Laboratory in 1947 at the company's Physics Training Program. After a year at the Schenectady Research Laboratory, he took his last assignment at the Lamp Division Research Laboratory at Nela Park and remained there for four more years. In 1952, he joined the Westinghouse Lamp Division Research Laboratory in Bloomfield, NJ. The next nine years were devoted to the investigation of low pressure discharges, principally mercury rare gas and fluorescent lamps. In 1961, he began work on the high pressure mercury discharge and the addition of metal halides to it. The following year, work on the high pressure sodium lamp was included and for the next 12 years, effort was divided between metal halide, high pressure sodium and other alkali metals. Much of the work was under government contract for laser pump sources, underwater lighting and color- adjustable sources for camouflage. Beginning in 1974 and continuing to his retirement in 1983, his full time was devoted to high pressure sodium lamp development and the solution of problems in the operation and production of the lamp. Mr. Larson is the author or co-author of 14 papers and has been granted 30 patents for his work.

Dr. Doug Mattox

After taking BS and Ph.D. degrees from Rutgers University in 1965, Dr. Mattox joined PPG's Glass Research Center in Harmarville, PA. After two years, he moved to the Westinghouse R&D Center in Pittsburgh, PA where he worked on new glasses for incandescent lamps, diffusive coatings and ITO deposition. In 1971, he became manager of a research group focused on lamp coatings, new HPS arc tube materials and fused quartz processing. After Westinghouse divested its lamp business, Dr. Mattox applied his glass interests to electronic packaging problems for Westinghouse's Defense Center and in 1989 came to teach ceramic engineering at the University of Missouri-Rolla. He continues to be active in research and industrial consulting in glass and coatings.

Dr. Georg E. Marsen

Dr. Marsen joined Osram GmbH in 1960 and worked in managerial positions in discharge lamps, GLS lamps and lamp components production. Key positions followed with responsibilities for discharge lamp development and technical operation in Osram Ltd. UK. After 10 years in the United Kingdom, he is currently head of the Osram prematerial and components factory at Schwabmuenchen, Germany.

Dr. David M. Moon

Dr. Moon is Manager of Advanced Material Technology at the

Westinghouse Science and Technology Center, where he has served since 1965. In addition to his work on doped tungsten, he has been involved in a broad range of material development programs focused mainly in the power generation field. He is the author of more than 40 publications and is a Fellow of ASM International. Dr. Moon received his BS degree from Michigan Tech and Ph.D. from Carnegie Tech in 1965.

Erhard Rasch

Mr. Rasch is a physicist and since 1957 has served the lamp industry with Osram, starting in research with glow lamps. Currently, he is head of the development of electronic ballasts. From 1965 to 1977, he was responsible for the development of fluorescent lamps and starters. Later, he worked on the development of lamp systems. During this research, Mr. Rasch focused on the development and introduction of compact fluorescent lamps with integrated ballasts.

Steven W. Sirek

Mr. Sirek's career includes nine years with GTE Sylvania (now Osram Sylvania) and 12 years with Philips Lighting Company (which purchased the Westinghouse Lamp Division in 1983). Most of his career has been in product development and factory support for tungsten halogen and infrared lamp manufacturing. He has worked with the product and in management positions at both companies. He has served on ANSI committees regarding halogen lamp standards. Most recently, he worked as a consultant until taking a full-time position in the aerospace industry. Mr. Sirek received his bachelor's degree in mathematics and physics in 1967, and his master's degree in physics in 1973.

Introduction

This book is about lamps which have been developed during the years following the end of World War II to the present. It was a period which can rightly be classified as the "Golden Age" of lamp design and commercialization. The U.S. lamp industry entered this period practically unaffected by the ravages of the war, in contrast to those of its major foreign competitors, notably in Germany, Holland, England and Japan where much of the lamp manufacturing facilities had either been destroyed or rendered inefficient for lack of spare parts and poor maintenance. Later, however, an event occurred which prevented the U.S. lamp industry from taking full advantage of its unique situation for a number of years. It was the ruling in an antitrust suit brought by the U.S. Justice Department against the General Electric Company, Westinghouse and a number of smaller lamp companies in 1942, that forced the lamp industry of the United States into a dramatic restructuring.[1] This ruling ended the monopoly that GE, and to a lesser extent, Westinghouse, had held over the U.S. lamp industry during the first half of the century through license agreements linked with production quotas. The consequence: "The lamp companies had to go it alone, a step for which they were not prepared with the exception of General Electric."

GE's licensing and lamp manufacturing quotas, which it imposed upon the other lamp manufacturers, had assured General Electric of supremacy over the U.S. lamp industry. Westinghouse was granted about one-fourth of GE's lamp sales, while the smaller lamp companies, in the aggregate, were permitted about 10% of the total. For paying royalties and accepting sales quotas, GE, in turn, shared its lamp designs and its manufacturing technology with the other lamp manufacturers. The com-

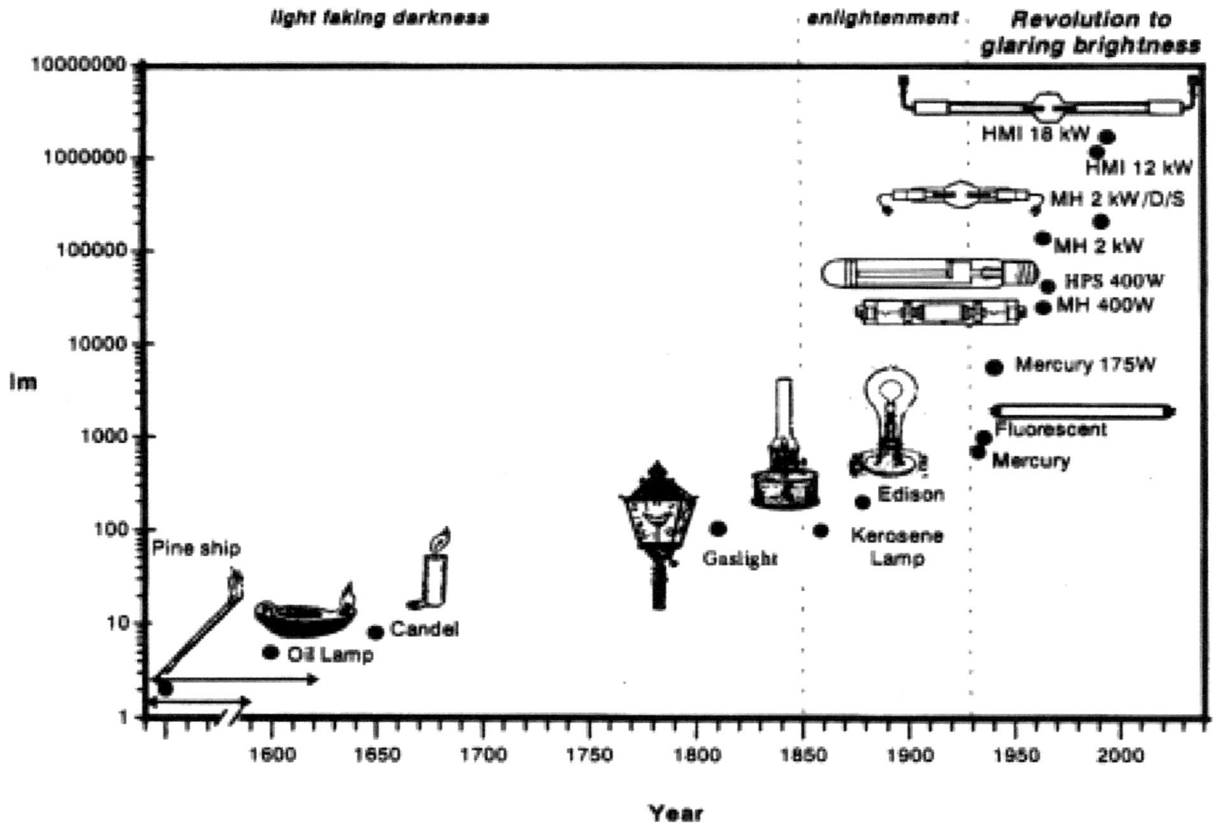

Figure 1. Luminous flux of various light sources during the centuries.

mercial interests of GE thus not only controlled the development of new light sources, it even impeded it.[1]

Freed from those constraints, all U.S. lamp manufacturers were propelled into fierce competition for lamp markets, and to meet these challenges, some, like Westinghouse, built up their own lamp and equipment engineering facilities. At the same time, the foreign competition, notably Osram in West Germany and Philips in Holland, rapidly rebuilt and expanded their plants and engineering laboratories and soon made their presence felt throughout the world. The evolving global competition was further intensified after 1973 as a consequence of the oil embargo, which made energy conservation a worldwide goal. Lighting, due to its high visibility, became an immediate target for energy conservation measures. This added urgency to the quest for the development of more efficient lamps.

Looking back to 1950, electric indoor and outdoor lighting was then primarily by incandescent lamps with their inherent good source color but inefficient conversion of electrical energy into light emission, commonly expressed by the ratio of lumens per watt. Fluorescent lamps of the 40W T12 type, using white phosphors, had started to make strong inroads into commercial lighting applications such as offices, warehouses and hallways. High pressure mercury lamps had made their debut in street lighting as early as 1940, but were not yet a major factor in outdoor illumination and not at all in indoor lighting. Although fluorescent lamps as well as high pressure mercury lamps were significantly more efficient than incandescent lamps and offered much longer service life, they left much to be desired with respect to source color and their color rendering quality. Nevertheless, these lamps, as well as incandescent lamps, provided the basis for the development of a spectacular variety of new lamps, including lamps that designers had only dreamed about in earlier times.

It would, therefore, seem appropriate to leave a record of the lamp industry's accomplishments.

While under the U.S. space program, NASA developed the technology for the sending of lunar landing modules and astronauts to the moon, the lamp industry perfected the fluorescent lamp and the high pressure (HP) mercury lamp. Lamp efficiencies and source color were greatly improved through painstaking research into new phosphors, phosphor blends and phosphor and lamp processing. The fluorescent lamp line was expanded with smaller and longer lamps and by bending lamps into U-shapes and spirals. The commercial impact was tremendous, as it opened many new lighting opportunities in the commercial, industrial and even residential markets. Similarly, high pressure mercury lamps were color improved with newly developed phosphor coatings of the outer lamp envelope, and the lamp line was expanded from 40 to 1500 watts. For a considerable time, color improved high pressure mercury lamps illuminated highways, parking lots and sports arenas.

Not to be neglected, the "bread and butter" incandescent lamps of the industry were given much attention in those days. Silica-coated lamps provided a more pleasing light. Lamp blackening, or the "water cycle," a cause for significant light loss during life, was attacked by employing better gettering materials. Overall, lamp quality gained much through the development of high-purity doped tungsten wire used in filaments. More was in store: The introduction of tungsten halogen (TH) lamps. To eliminate wall blackening caused by evaporation of tungsten from the filament, double-ended TH lamps consisted of a quartz tube with axially mounted filament. Initially in the development, the lamps were dosed with solid iodine, but later mostly with gaseous bromine compounds. These TH lamps became very popular in the outdoor illumination market. The high temperature required to facilitate the reaction of the halogen with the tungsten necessitated the use of quartz glass for the lamp envelope. The original lamps spawned the development of a wide variety of lower-wattage, often miniaturized, TH lamps which made major inroads into the commercial market. However, TH lamps found their most spectacular application in automobile headlights. In many instances, the industry succeeded to replace the expensive quartz envelopes with identical ones of borosilicate (hard) glass. It was not to be the end in the development of incandescent lamps. From outdoor illumination with high-wattage lamps, this line of lamps was miniaturized and now serves in low-voltage and low-wattage versions for merchandise lighting applications.

The culmination in the development of extraordinary light sources occurred in the early '60s with the introduction of high pressure metal halide (MH) and high pressure sodium (HPS) lamps. Building on the high pressure mercury lamp design, adding halides to the mercury discharge greatly increased the efficacy of the HP mercury lamp and also improved its source color and color rendering abilities. The initial versions were of the high-wattage type which yielded efficacies of 80 to 130 lumens per watt. Newer versions of this lamp type are designed for wattages as low as 35W. Incredible as it may appear, metal halide lamps are now used for automobile head lights.

The most intriguing new lamp, without any equal, is the HPS lamp. With the exception of the low pressure sodium lamp, the HPS lamp exceeds any other lamp type in luminous efficacy, and matches most other lamps in service life. This lamp's coming into being was made possible by the introduction of translucent sintered alumina. The high pressure in the arc tube causes line broadening of the sodium to the extent that, in general, lamp color is quite satisfactory and can even be adjusted by design to provide remarkably good color. Recently developed low-wattage miniature versions, in combination with MH lamp additives, hold great promise for residential use. At present, HPS lamps have become the preferred light source for roadway, street, parking lot, warehouse and some industrial lighting.

Electric lighting had its beginning barely 100 years ago, when Edison patented his commercial carbon filament incandescent lamp in 1880.[2] Edison's success in perfecting the carbon filament incandescent lamp was significant in giving direction to the development of lamps based on other physical principles rather than thermal radiation, notably on the conversion of electrical energy into light emission through means of gaseous discharge or high frequency operation. He demonstrated that success, in large measure, depended on material breakthroughs. The achievements in lighting during the second half of this century fully proved this point as they would not have been possible without major advances in lamp construction materials. Testimony to this fact are the production of high-purity doped tungsten wire for filaments and electrodes, the development of unique and more efficient phosphors, the advanced technology for the production of high-purity quartz tubing to be used for TH lamps as well as for MH lamps, and the development and production of translucent polycrystalline alumina (PCA) tubing for HPS lamps.

One other point deserves to be made. The revolution in lamps owes its success to two other important factors. One is the commitment of the management of the lamp companies to provide long-range financial support for the development projects it considered necessary for it to remain a contender in the lamp markets. The second factor is the long-term career commitment of a dedicated and relatively small group of scientists and engineers who devoted their know-how and experience to the development of these lamps and made "The Revolution in Lamps" possible. Some of them have contributed to this book.

References

1 Arthur A. Bright, Jr. The Electric-Lamp Industry: Technological Change and Economic Development from 1800 to 1947. New York: The MacMillan Co., 1949.

2 Edison, Thomas Alva: "Electric Lamp," U.S. Patent No. 223,898, issued Jan. 27, 1880.

Chapter 1

The Incandescent Lamp

by Chester Dawson and Douglas Mattox

Thomas Alva Edison did not invent the incandescent lamp—he commercialized it. How to make it affordably was Edison's contribution and he did not lack competitors. The incandescent lamp represented such a remarkable aggregation of sophisticated technologies—from vacuums to refractory filaments to hermetic sealing—that there was enormous room for improvement. In poring over the early literature, none of the inventors can be found resting on his laurels. An extraordinary animation is evidently driving them to perfect this extraordinary product. This zeal appears to be contagious since an uninterrupted concern to improve this device can be witnessed in the 100+ years of its existence.

To the inventors, technologists and engineers, every direction of interest represented unlimited horizons for improvement. To consider these contributions, the incandescent lamp must be examined in its component parts and the interrelationship between these parts. In the following sections, an uncompromising commitment to improvement and pride of accomplishment is evident on the part of the participants.

Lamp Chemistry

Vacuum Lamps

Edison recognized the importance of creating as high a vacuum as possible in his first laboratory-produced lamps. Ten hours before beginning their famous 40-hour life test watch on October 19, 1879, Edison and Francis Jehl began the tedious task of evacuating the envelope, which history would acclaim as the first truly successful incandescent lamp. Jehl would later note on the events of that day:

> Edison sitting in a chair directly opposite the pump, watched as it worked, noting the large cylinders of air being pressed down by the likewise large cylinders of mercury ... During the whole operation I was kept busy transferring the mercury as it passed out of the glass tube of the pump into the jar. To do this I substituted an empty jar and mounting a small stepladder, poured the full one into the reservoir on the top of the pump stand, from which the mercury was fed and regulated in the pump by a rubber hose ... Edison continued to apply the battery current to the carbon lamp filament from time to time, increasing its intensity as well as the time of its application until the occluded gases were driven out.[1]

It has long been known that impurities within the sealed glass envelope of an incandescent lamp can have disastrous results, not only in shortening the expected life of the lamp but also in prematurely and sometimes severely blackening the bulb wall. The first commercial Edison lamps were vacuum lamps, exhausted using Sprengel mercury pumps[2]. The goal, of course, was the unattainable perfect vacuum. Howell and Schroeder[2] report that, in June 1881, it took five hours to evacuate a lamp with each exhaust operator handling about 50 lamps at a time. Today a single automated seal and exhaust machine can handle up to 10,000 lamps per hour.

One of the primary problems with obtaining that perfect vacuum was that gases, in particular water vapor, would attach themselves to all the lamp parts. Lamp engineers soon discovered that heat was required to remove the water vapor from the various surface within the bulb. Unfortunately, mercury displacement pumps did not pump water vapor very well, so a cup of phosphoric anhydride was attached to the mercury pump to absorb the water vapor. A breakthrough finally occurred in 1894[2] when an Italian, Arturo Malignani, discovered that red phosphorus, painted onto the inside of the exhaust tube and vaporized during the exhaust process, vastly improved the vacuum inside the lamp and, in turn, lamp performance. John Marshall would later show that the lamp filament could be dipped in the red phosphorous compound and then when the lamp was lighted or flashed the phosphorous would burn off the filament, reacting with gas impurities as it did so.

The next significant breakthrough in the gettering or removal of unwanted gas impurities in vacuum lamps involved halogen gases and salts. One of these, a double halogen salt called cryolite (Na_3AlF_6), was invented by a GE scientist Harry Needham.[2] Properly applied, it became a very effective getter for vacuum lamps. When flashed or burned off the filament at light up, the cryolite coats the bulb wall and reacts with the tungsten vaporizing from the burning filament. There it forms a light colored deposit rather than black, reducing lamp blackening during use.

Gas-filled Lamps

These early improvements on the basic Edison lamp were aimed at vacuum lamps. Many scientists investigated inserting inert gases into the lamp, but with little success. The biggest problem was that the inert gas, usually nitrogen, would drastically reduce the filament temperature and in turn the light output. Irving Langmuir then found that, while it was true that the heat loss from small, low-wattage filaments in gas filled lamps was too great to make them practical, heat loss did not increase proportionally as the wattage and filament size increased. Filling the very high-wattage lamps (2,000W and up) with inert gases would not drastically reduce their light output; but this was not the case for the more common 40-150W general service lamps.

Then came the discovery for which Langmuir will be forever known. He found that the heat loss of the filament was not determined by the diameter of the wire from which the filament is made, but by the diameter of the wire wound filament or coil itself. Coiling the filament, therefore, would increase its effective diameter, allowing a 60W or 100W lamp filled with inert gas to burn efficiently. The gas layer surrounding the filament has now become known as the Langmuir Layer.

Langmuir's work did not lead to the immediate development of practical gas-filled incandescent lamps. Trying various gases of different pressures, it was found that a mixture of about 85% Ar and 15% N_2 at slightly under atmospheric pressure worked well. The pressure had nothing to do with lamp chemistry. Pressures in excess of atmospheric pressure were considered too dangerous; lamps could explode. Lower pressures were not as effective at improving lamp efficacy nor in providing the best combination of life and light output. Argon worked well in that it was an inert gas and did not conduct heat away from the filament as well as other gases. Early in this century, however, the technology did not exist to obtain sufficient argon gas to make millions of lamps, and when used by itself, arcing across the filament would occur. By adding 15% N_2, the arcing problem ceased and the development of means to obtain argon in large quantities resolved the remaining impediment. During the 1970s, the availability of krypton gas, which has lower heat conductivity than argon, gave lamp designers another tool to add efficiency to the incandescent lamp.

The need for purity in gas-filled lamps is just as critical as eliminating residual gases in vacuum lamps. Evacuating and then flushing with nitrogen several times—while at the same time baking the bulb wall prior to finally filling the lamp with gas—results in a very pure lamp atmosphere. Even then, phosphorous getter is still needed to remove remaining impurities.

The development of the silica-coated lamp in 1948 and the need for smaller and hotter operating bulbs resulted in a resurgence of research on getters and lamp chemistry. At Westinghouse, that effort was led by a devoted scientist, Earl Gulbransen. Dr. Gulbransen would continue his work for more than 30 years.

The "Water" Cycle

One area that generated a tremendous amount of study at Westinghouse as well as most other manufacturers was how to arrest what has generally become known as the "water cycle." When water vapor comes in contact with a hot filament, it dissociates into H_2 and O_2. The O_2 may then react with the filament to form one of many tungsten oxides depending on the temperature where the reaction occurs and the partial pressure of the oxygen. The tungsten oxide formed is volatile at these temperatures and deposits on a cooler surface, which is usually the leads or bulb wall. At the correct temperature, the hydrogen, which previously dissociated from the water vapor, reduces the tungsten oxide to H_2O and tungsten, leaving the tungsten as a deposit on the leads, the cooler portion of the filament or bulb wall, and also leaving behind the H_2O available to repeat the process all over again. Figure 1-1 and Figure 1-2 show examples of deposits forming on the lamp leads and filament when the water cycle is severe. In Figure 1-1, the tungsten is forming whiskers growing out from the lead, while in Figure 1-2, crystals are growing on the filament close to where it clamps to the lead.

The water cycle is not as "ideal" as first thought and the cycle decreases with time or may stop altogether if the water partial pressure

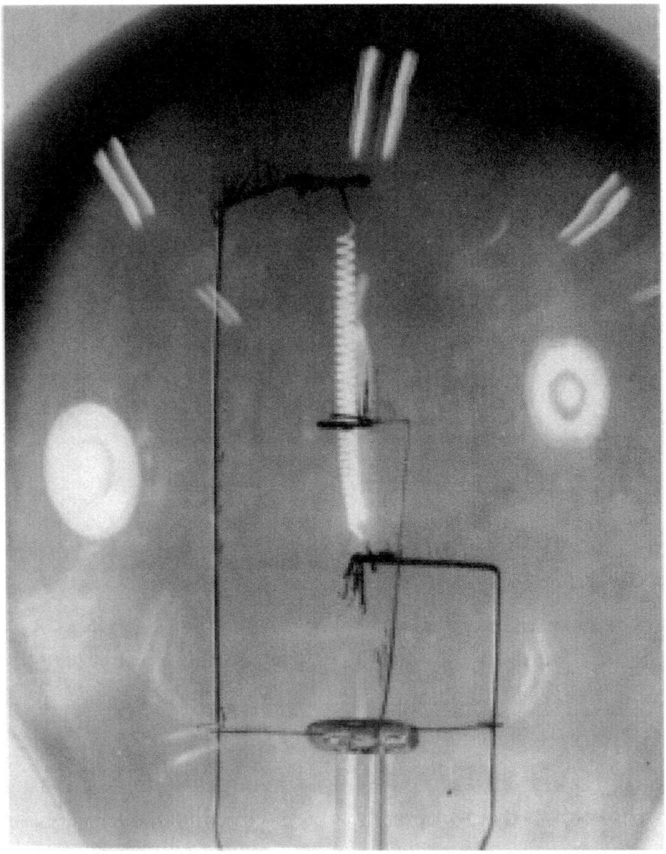

Figure 1-1. Example of severe "water cycle," where WxOy whiskers have formed on the leads and center support of a 100W lamp. Convection currents within the burning lamp cause the whiskers to stand on end. Courtesy of Philips Lighting.

drops low enough. Ample direct mass spectrometric evidence[3] shows that tungsten oxides are only produced at relatively high pressures of oxygen. At lower levels, oxygen desorbs from the hot tungsten either as an activated molecule or as atomic oxygen, leaving the tungsten intact. Scientists believe that the same processes occur with water in place of oxygen except that hydrogen also is produced. Thus, as long as the desorption of water from the lamp parts occurs slowly enough to keep

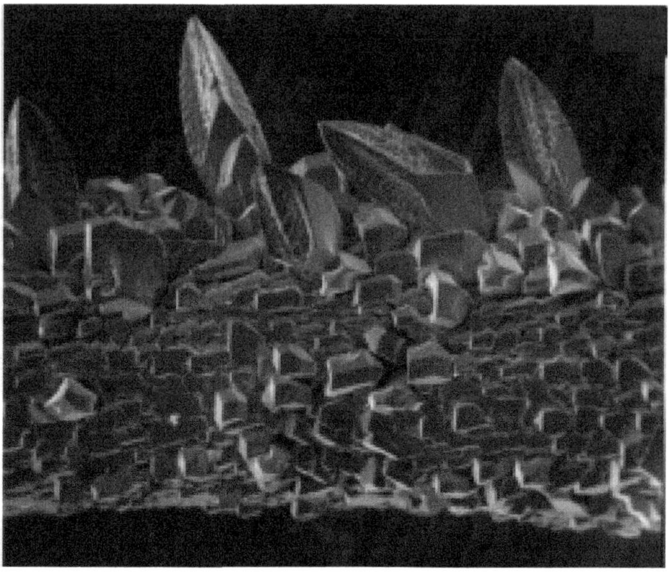

Figure 1-2. As a result of the "water cycle," tungsten crystals have grown near the cooler, clamped end of a 100W lamp filament. Courtesy of Philips Lighting.

the partial pressure of water low, maintenance getters adequately remove the oxygen and hydrogen from the lamp before these gases cause filament erosion.

Maintenance Getters

As early as 1950, Dr. Gulbransen along with Ken Andrew tried to stop the water cycle by removing the H_2 to prevent the cycle from continuing. This was particularly important for small hot burning automobile lamps such as the 1157 stop/tail lamp. Since the early 1930s, many well-known scientists—including G. Hagg, D.P. Smith, J.H. deBoer and J.D. Fast[4]—studied zirconium as a maintenance getter for incandescent lamps. These studies showed the reaction between the zirconium and hydrogen varied tremendously depending on the preparation of the zirconium getter and the temperature of the reaction. They did not address the rate of the reaction and the influence on this rate by the co-reaction with oxygen and nitrogen. Gulbransen and Andrews set out to obtain a basic scientific understanding of these reactions.

Gulbransen found that the oxide films forming on the Zr getter limit its rate of reaction with H_2 and the rate of reaction varies tremendously with how the oxide film formed.[5] For example, oxide film layers forming at room temperature were more resistant to H_2 gettering than were much thicker film layers formed at higher temperatures. Today, we know that Zr getters work best if they are prebaked at temperatures up to 1275K.[6] Since these temperatures would destroy most lamps, we have placed the getter where the radiation from the filament and conduction of heat within the lamp can heat it to between 650K and 1100K. As the temperature of the getter increases, the oxide layer begins to diffuse into the metal. Sustained gettering of oxygen occurs at temperatures above 975K where the oxide goes into solid solution. Such temperatures are therefore required for the practical application of Zr as a continuous getter. Consequently, the oxygen-layer thickness is a balance between the rate of diffusion and the rate of adsorption of additional oxygen. Simultaneously, gettering of hydrogen may occur but at any temperature above approximately 775K, the quantity of hydrogen that can be gettered by Zr is limited, mainly to the solid solution phase as dissociation of the hydride becomes significant.[7]

A Swedish scientist, Sven Jansen, joined with Dr. Gulbransen as he continued his thermochemical analysis of maintenance getters. By 1967, Gulbransen and Jansen would complete a thorough study of the interaction of H_2 and H_2O at various partial pressures and varying bulb wall temperatures using both Zr and Zr alloy getters. By the early 1970s, funds at Westinghouse for basic research on incandescent lamps were being diverted to newer lamp technologies. Dr. Gulbransen's efforts began to move from basic research to more applied research. He joined with John Condron to examine the reactivity of numerous gas impurities including CO, CO_2 and methane with Zr, and Zr-Al getters and how these reactions affected the adherence of the getter to the lamp leads.

Hot Can Testing

At this time, most lamp manufacturers were using some form of hot can test to determine how well the lamp was baked out during manufacturing and how well the maintenance getters were performing. These cans varied from two-quart fruit juice cans to aluminum beer cans to cans with holes or blackening inside, outside or both and so on to conventional top-hat type fixtures. Beer cans were most commonly used by Westinghouse to test general service 60-100W silica-coated lamps.

Bulb walls of lamps burned in these cans reached temperatures approaching 625K. At these temperatures, significant out-gassing would occur if the lamp was not processed properly during manufacturing. Varying the can size or blackening them controlled the bulb wall temperature. This would lead to new insights into how getters work by two more Westinghouse scientists: Dr. Jack Singleton and Dr. Al Pebler.

Singleton developed a technique to attach one end of a tube to the exhaust tube of a finished commercial lamp and the other end to a gas-analysis mass spectrometer. Cracking the exhaust tube inside the tube connected to the mass spectrometer allows the lamp to be pumped directly to the mass spectrometer and any gas impurities studied. During this same period, similar work was performed by Pebler, who teamed up with Ken Andrew to use mass spectrometry to examine red phosphorous, P_3N_5 and these in combination with Zr / Al maintenance getters for in situ experiments on many different lamps.

Positioning of Getters

Soon, Singleton began working with another scientist, Chet Dawson. Their work would lead to optimizing the application and positioning of maintenance getters inside the lamp and into using basic laws of radiation combined with some empirical experiments to design a theoretical mathematical model for the design of the lehrs used to bake out bulbs prior to exhaust and sealing.

As the 1970s came to an end, research on incandescent lamps at Westinghouse would shrink significantly. Work by Gulbransen, Jansen, Pebler, Andrews, Singleton, Dawson and others greatly increased the overall understanding of gettering in both vacuum and gas-filled lamps. Today's lamp engineers routinely use that knowledge to ensure the proper selection and location of getters which, in combination with the correct processing, provides the customer with the best quality of product available.

Mechanisms of Lamp Failure

Prior to the incandescent lamp, when the light failed it did so with some warning—i.e., it flickered because a flame extinguished or the fuel supply became unreliable. But the incandescent lamp failed unexpectedly, suddenly and with a subconscious insult to its user—it deprived him of light and caused him unplanned effort and expense. Although all lamps burn out, it always annoys and surprises the user. Incandescent lamps do not fade out, and as lighting proliferated, so did the number of failures (and they were more memorable than the lamp's service). This proliferation, in spite of the continuous improvements in output and efficiency, led users from the '50s to be more conscious of failures and eventually to volunteer that "lamps weren't as good as they used to be." The common story told into the '60s was of the old house purchased with a 50-year-old lamp in some remote location which still was operational. The fact that the lamp had very low lumen output by modern standards, indicative of a cool filament, was too esoteric an explanation for the mass user. Manufacturers wanted to find ways of improving quality, in particular, reproducibility. To do this, they would have to narrow the natural distribution of lifetimes experienced by lamps. This would particularly mean reducing early failures.

Light Output Versus Life

Incandescent lamp design is a give-and-take between light output and lamp life. At its 3650K melting temperature, a tungsten wire generates 52 lumens per watt (LPW) of light but has zero expected life. To obtain its designed lifetime of 750 hours, a typical 100W coiled coil filament will burn at about 2860K and because of radiation and heat losses generate about 17 LPW instead of the theoretical maximum efficacy of 21 LPW.[7]

Figure 1-3[8] indicates the relationship between lamp life and luminous efficacy. These relationships are the statistical average for a large number of lamps if they fail by normal means. Lamp manufacturers generally determine expected lifetimes by burning lamps on open racks during which they periodically switch the lamp off and on. In general, these expected lifetimes do approximate what is realized by the end user; however, what exactly causes a lamp to fail is not clearly understood and led to significant research during the '60s and '70s.

Failure by Chemical Attack

The previous section discussed the need for clean atmospheres and getters in lamps to prevent premature blackening and filament failure. This type of failure is caused by chemical attack, which often occurs near where the filament is clamped to the lead. At this point, the gas flow and

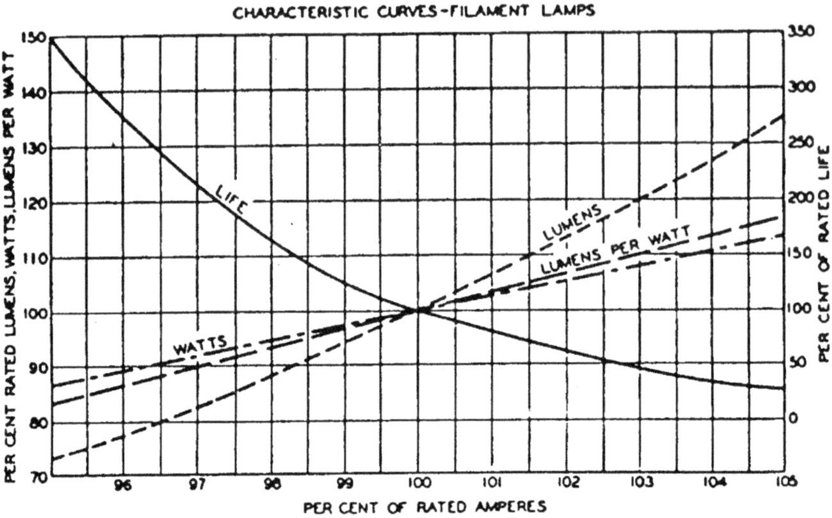

Figure 1-3. Relationship between lamp life and light output (lumens) for general service incandescent lamps. Circa 1980 *Westinghouse Lighting Handbook.*

filament temperature are most ideal for prevalent water to react with the filament. This often occurs when using lamps, inadequately baked during manufacture, in very hot fixtures such as recessed downlights. The resulting lives would be shorter than expected. Generally lamps that have a crack or leak fail immediately when turned on. These types of failures occur from improper manufacture or handling, but how do lamps fail when made to specification?

Influence of Hot Spots on Lamp Life

The basic principles of incandescent lamp design predict that the rate of evaporation from the filament determines lifetime. Temperature accelerates the evaporation and thus shortens the life while higher gas pressures slow evaporation and prolong the life. If the evaporation were to occur uniformly over the entire filament, the resistance of the wire would increase as the wire thinned, resulting in a corresponding drop in temperature. Under these ideal conditions, the lamp would last forever. Instead, the wire burns through at a specific spot, while the remainder shows little or no reduction in diameter. These spots are caused by some localized defect and have become known as hot spots. Becker in 1925 was the first to suggest a hot spot model of filament failure.[9] In his model, the lamp filament would fail when a portion (hot spot), which was initially hotter than any other section, reaches the melting point of the wire. The time to failure depended on the initial temperature of the wire and the temperature difference of the hot spot. Horster et al. were among the first to try to put a quantitative interpretation to the effect hot spots have on the lifetime of incandescent filaments.[10] They were able to demonstrate a relationship between the smoothness, or lack of hot spots, along a tungsten wire and its weight loss at failure. The smoother or more uniform the wire, the more uniform the evaporation will be, and the greater its weight loss at failure.

The '70s brought on considerable efforts, particularly at General Electric and Westinghouse, to further understand the hot spot phenomena.

At Westinghouse, filament failure research was being conducted by a number of scientists, including Fran Harvey, George Comenetz, Herman Johansen, Al Pebler, Heinz Sell and Chet Dawson. While Dr. Harvey was developing a theoretical model for hot spot growth,[11] Dr. Comenetz was showing that the necking down or thinning of the filament wire effected the life of coiled filaments as well as straight wires.[12] Because it is thin, the wire's cross-section at that point is of higher resistance than at other points along the wire. Rather than have an overall drop in current, which would occur if the entire wire thinned, the current is hardly affected by localized thinning. More energy is dissipated at this point then elsewhere, causing the wire to burn hotter or form a hot spot. Dr. Harvey's and Dr. Comenetz's model, like those before them[9,10] would predict that because of the higher temperature, thinning would accelerate at the hot spot until the lamp eventually failed at this point.

Effect of Coil Pitch

When lamp manufacturers wind tungsten wire into either a single coil or that coil into a second coil to form respectively a single coil or coiled coil filament, the distance between each turn of the coil is very uniform; when mounted into a lamp, however, that coil is stretched, threaded through and pulled by supports and finally flashed (lighted briefly) to evaporate the getter and set the metallurgical structure of the wire. While the manufacturer tries to use processes to minimize changes in coil spacing (called coil pitch), variations do occur. Al Pebler used x-

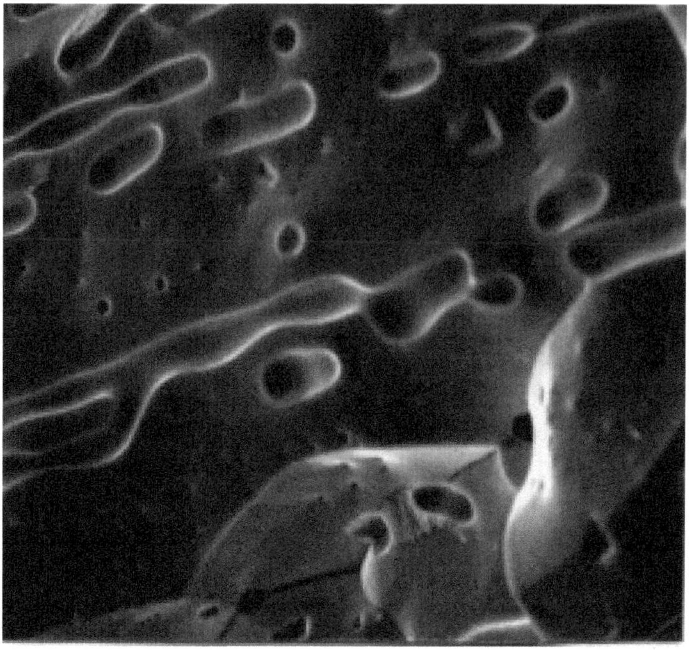

Figure 1-4. Scanning electron micrograph of a longitudinal fracture of an annealed tungsten wire previously ion implanted with potassium and aluminum. Shown are metastable elongated pores beginning to spheroidize into the more stable rows of bubbles typically seen in lamp-grade annealed tungsten wire. Courtesy of Philips Lighting.

ray radiographs to correlate this variation in coil pitch with the failure point in both coiled coil and single coil filaments. Filaments burn the hottest where the coil pitch has been decreased (turns become closer together) most during processing, and this is the point at which they will often burn out.

Influence of Internal Flaws

Thinning of wire and variations in coil pitch are two ways that hot spots form, but what if the wire's cross section diminishes by the development of internal discontinuities? In another section of this book, Sell and Moon discuss how lamp-grade tungsten wire is produced from tungsten powder doped with potassium silicate and aluminum chloride. The potassium dopant is the source of submicroscopic bubbles that form when the filament is heated. Figure 1-4 shows metastable, elongated

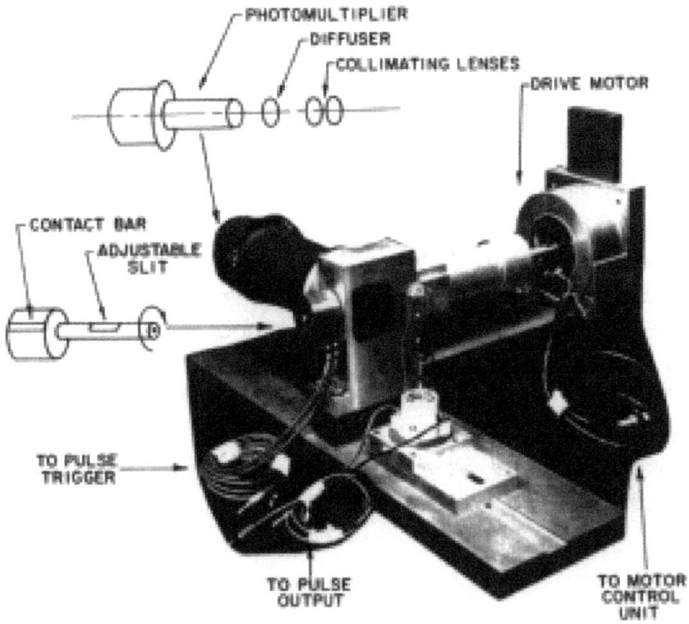

Figure 1-5. Current-pulse scanning apparatus used to track hot spots during the lifetimes of incandescent lamps. Courtesy of the Illuminating Engineering Society of North America.

pores in drawn, annealed tungsten wire, that was previously ion implanted with potassium and aluminum. The pores are seen beginning to spheroidize into the more stable rows of bubbles commonly seen in lamp grade annealed tungsten wire. Chet Dawson showed that thermal gradients, such as hot spots, act as a driving force for bubble growth.[13] Bubble migration up the temperature gradient is the most likely of several possible mechanisms that may be responsible for this growth. When two bubbles come in contact with one another they coalesce and form a larger bubble. The result of this phenomena would be thinning of the wire from within as well as from without with a shorter resulting lifetime than would be expected from the hot spot evaporation model alone.

Experimental Proof That Hot Spots Influence Lamp Life

Other than chemical attack, most incandescent lamps fail by the development of hot spots. Dawson et al., using a current pulse scanning device (see Figure 1-5), conclusively followed the development of hot spots in T10 single coil filament lamps from the beginning to the end of life.[14] By pulsing the filament, they traced the temperature profile of the filament when such temperature differences were greatest. This may also explain why lamps fail so often when they are switched on.

When a lamp is turned on, the temperature at a hot spot rapidly overshoots the steady state burning temperature of the filament. This occurs because thermal conductivity does not have time to level the temperature closer to the steady state burning temperature of the rest of the filament. Since the peak temperature a hot spot reaches on switching may be several hundred degrees higher than its steady state temperature, it will often fail when this temperature overshoot approaches the melting point of tungsten. This did occur in the work of Dawson et al. where the temperature overshoot at switching increases during lamp life until the lamp filament eventually fails at the fastest developing hot spot.[14]

The Glasses of the Incandescent Lamp

The First Lamp Glass

"We blew a bubble for a man named Edison (1877)," said a Corning Glass worker.[15] Those glass workers had no idea of the significance of their labors for this man who would become known as the "wizard of Menlo Park." Hand-worked glass was the order of the day for the first incandescent lamps and continued to be so until production volumes directed attention to this bottleneck (circa 1880).

Edison's original invention[16] exclusively focused on the practical means of producing an effective seal between the glass and the penetrating electrodes. Selection of glass was an afterthought and the choice delegated to the glass maker, Corning Glass Works. Thus, the original envelopes were free-blown from one-inch glass tubing, which produced the characteristic tip-off at the top of the early lamps. During the first 12 years of production, the envelopes were made from lead glass formulations (20-30 wt. % lead oxide). The first glasses were selected from those familiar to the glass workers in that day. Thus they chose lead crystal glasses which had excellent working characteristics. These glasses also possessed good electrical insulating characteristics which were required for the DC power applications of the day.

Beginning in 1915, soda-lime glasses (essentially bottle glass) began to replace the lead envelopes as automatic equipment was developed. The interior lamp parts were still made of glass similar to the original lead glass envelope compositions. This was due principally because of the excellent electrical performance of these glasses but also because the automatic assembly equipment, which would be developed, mimicked the hand work and their preferred glasses, the lead oxide glasses.

Lamp workers call lead oxide glasses "long" because their viscosities, or thicknesses, change slowly with temperature. This means the glass, once heated to an appropriate softness, can be manipulated dependably without having to attempt to regulate the temperature excessively. The thought of requiring precise temperature control would have intimidated the designers of future automated machinery since the typical high production assembly machine would comprise 160,000 parts by the '40s. The skill and art required to keep these machines reasonably productive created considerable resistance to changing any of the materials, particularly the glasses.

Through the '60s, the typical incandescent lamp consisted of three types of glass. Beginning in 1927, the envelope was made on high-speed ribbon machines from soda-lime silica glass. The flare was made from glass tubing consisting of approximately 20 wt. % lead oxide. Finally, the stem/button was produced from a 30 wt. % lead oxide glass. In spite of the presence of the three glasses, they were to behave indistinguishably once joined together. The lead oxide flare and stem constructions were ultimately transported to fluorescent lamp manufacture.

The Need for Change

Long before the present environmental concern to eliminate lead and its compounds from commercial products, the lamp industry wanted to eliminate it from incandescent manufacture for economic reasons. There have always been strong economic pressures on incandescent lamps because of the combination of demand, volume of production and the nature of retailing (super markets, discounters.) By the '60s, lead oxide represented as much as 80% of the batch cost in the high lead glass.

This cost penalty was even greater if the manufacturer was required to convert often between different lead content glasses. Production would continue for a time with the lower lead glass and then the tube furnaces would be converted over to the higher lead glass and the sequence repeated. This caused loss of glass during transition and excessive wear of the furnace refractories because of the conversion. The first line of attack on the unfavorable economics was to attempt to produce a single 20 wt. % lead glass which would serve both the needs of flare and filament.

In the mid-sixties, a three-year effort was mounted at Westinghouse to identify the essential function of the glass components and to duplicate the functionality of the 30 wt. % lead oxide glass in a 20 wt. % glass. Remarkably, even in the late '60s, there was a requirement to survive dc application requirements because federal contracts required all lamps to meet their most severe possible application specification (dc), regardless of use. As a major customer, their requirements could not be ignored. Thus, to meet an application far smaller than 1/1000 % of the market, all domestic producers used the electrolysis-stable, 30 wt. % lead glass as the stem/button glass. At Westinghouse, Doug Mattox, having recently come from PPG Industries Glass Research Center, was given the assignment of determining if a 20 wt. % lead glass could be formulated which would satisfy all the requirements of both lead glasses.

Subtleties in Lamp Glass Chemistry

Lead oxide levels as high as 30 wt. % allow excellent manipulation as hot glass. Glass chemists have always felt it was beneficial to use both soda ash and potash together in glass compositions. On a weight percentage basis, the typical 20 wt. % lead oxide glass consisted of equal weights of alkali, about seven percent each. The 30 wt. % glass tended to have about 8 wt. % K_2O and 4 wt. % Na_2O, for a total of 12 wt. % alkali. The lowered total level of flux in the 30 wt. % glass was compensated by the fluxing effect of the increased lead oxide. Table 1-1 shows the typical compositions throughout the industry in 1970. The K_2O dominated the Na_2O since the 19th century because of its greater "briliancy!"[17] (Remember that these glasses were not developed for lamps.) The sole application for the higher lead glass was the stem button, which reached the highest temperatures and which could be the site of shorting by electrolysis of the alkali in DC application; thus, the lowered alkali level.

Alkali Effects

Glass scientists discovered an anomaly in alkali glass compositions known as "the "mixed alkali effect."[18] If, on an atomic basis, glasses are compounded in identical proportions with either sodium or potassium oxides, their electrical and viscosity properties are virtually identical. That is, there is virtually no distinguishability between the potassium or sodium glasses. If potassium and sodium oxides are intermixed, however, distinctly nonlinear effects are observed. Departures from linearity of 10^3 to 10^6 are observed, being greatest at equimolar proportions. Of most consequence to stem/button application, the electrical resistivity can increase by using equimolar fractions of alkali. It was recognized that this electrical resistive effect was not maximized according to the mixed alkali effect in either of the two classic lead oxide compositions. This appeared to present the opportunity to enhance the resistivity and consequent electrolysis by simply changing the alkali ratio.

Table 1-1 indicates that on a molar basis, neither glass was taking the benefit of the mixed-alkali, electrical resistance enhancement. (Totals in Table 1-1 do not add to 100 because minor oxides have been ignored for simplicity's sake.)

Table 1-1. Average proportions of lead glasses, pre-1970.

Oxide	*20 % PbO (weight %)*	*30 % PbO (weight %)*	*20 % PbO (mole %)*	*30 % PbO (mole %)*
Na_2O	7.0	3.6	9.0	4.7
K_2O	7.5	9.0	5.3	7.6
PbO	22.1	31.5	7.6	11.4
SiO_2	61.5	54.5	76.5	75.0
Al_2O_3	1.9	1.4	1.2	1.1

The New Single Lead Glass

In systematic measurements of experimental glasses, Mattox was unable to demonstrate that the mixed alkali effect could be observed in the lead oxide-based glasses. Instead, the resistivity and polarization tended to be purely proportional to the K_2O content. Thus a new 20 wt. % glass was identified which had about 11 wt. % K_2O and 4 wt. % Na_2O as the alkali content. This provided the required resistivity and viscosity characteristics previously provided by the 30 wt. % glass.

Attempts to Improve Light Diffusing Coatings

Ever since the first commercial lamp, attempts have been made to reduce the glare from the incandescent filament. As the efficacy of the lamps improved, from less than 2 lumens per watt for the first Edison carbon filament to several times that for Langmuir's coiled tungsten filament, the need for a light-diffusing coating became more apparent. Early attempts to reduce glare involved using white or opal glass or, sandblasting, acid etching or painting of the outside of the bulb. White and opal glasses were generally expensive and resulted in excessive light absorption, while a frost or paint on the outside of the bulb tended to catch dirt and also reduced light output.

Inside Frosting

Inside frosting was not thought to be commercially feasible because the single dip acid-etching process used in the early part of this century caused the formation of sharp angular crevices that weakened the glass. A 1912 German trade paper and 1918 research by R.W. Wood suggested successive acid etches to reduce the sharp edged crevices.[19] It was not until 1925 that these ideas were put to practical use.[2,20,21] At that time, Marvin Pipkin of General Electric successfully demonstrated the feasibility of inside frosting using a two bath hydrofluoric acid treatment.[19] Corning Glass was licensed to use the process and it remained as the only widely used method to obtain light diffusion until December 1948, when General Electric announced the "Q" Bulb.

Better Light Diffusion by Silica Coating

Marvin Pipkin claimed that the "Q" lamp, containing a coating of silica powder on the inside wall, had much better light diffusion characteristics than inside frost lamps with little or no sacrifice in lumen output.[22,23,24] In July 1949, Westinghouse introduced its own silica-coated lamp. A wet flush coating process was installed at its Trenton, NJ lamp plant and production of flush silica coated lamps quickly reached 200,000 a month.

Despite being much more expensive than the Pipkin process, Westinghouse originally used flush-coating because it was the simplest to put into limited production. Some manufacturers continued to use it into the

'80s. One of the alternative processes proposed was the electrostatic coating of fine amorphous silica particles onto the inside bulb wall. Westinghouse first demonstrated this method in April 1949.[25] Its obvious advantages were low material and labor costs and the possibility of attaching a coating unit directly to the lamp assembly equipment.

By April 1953, an electrostatic coating program under the direction of G. Meister and others at the Westinghouse Lamp Division's Research Lab along with the help of an innovative equipment designer, Stan Lopenski, had advanced to the point where a system was ready for a trial run. Within a year, Westinghouse was producing and marketing electrostatically coated lamps.

The successful and quick development of electrostatic silica coating can be credited to several Westinghouse scientists, but George Meister and Gaines Young along with the previously mentioned machine designer, Stan Lopenski, deserve the most credit. Meister and Young described their process for electrostatically coating incandescent lamps in January of 1950:

> In general, the method of coating is as follows: The bulb is placed on a rotating head and is heated by a flame (radiant burner). The heat renders the glass more conductive electrically, and the hot-ionized gases and radiant burner serve as one of the electrodes. The rotating head contains a nozzle supporting the probe which serves as the other electrode. The probe is a straight metal rod, pointed at the end in order to provide ample ionization current when the DC voltage is applied. The smoke, a mixture of air and finely suspended silica particles, comes from the smoke generator, goes through the expansion chamber where some of the heavier particles are lost, and is blown into the bulb through the nozzle. The smoke is charged by collision with the ions, and subsequently the charged silica particles are drawn over to the glass wall by the electric field and deposited there.[25,26]

Early Attempts to Improve the Silica Coating Process

Meister and his co-workers spent the next several years trying to improve the adherence of the silica powder to the glass bulb and trying to eliminate any residual water left in the powder. A small amount of water left in a sealed lamp reacts with the tungsten filament and causes the lamp to blacken and fail prematurely. Meister felt that the silica entering the bulb should be as dry as possible. The first lamps were electrostatically coated with moderately fine silica powder called Hi-Sil fired for several hours at temperatures approaching 1175K. With only a dust cover for protection, Hi-Sil fired in this manner remained dry for up to a week.[26] Unfortunately, the coatability of this powder diminished with age and it never did coat as well as unfired silica. Today we know that firing silica at temperatures approaching 1175K causes it to lose many of the surface hydroxyls to which water attaches and also initiates surface diffusion and sintering of the silica particles that diminishes its electrostatic coat-ability.

Another technique tried was "micronizing" (very fine grinding) the powder before coating and then steaming it after it was first coated onto the interior of the bulb. While micronizing became a mainstay of the silica coating process, steaming proved to be undesirable. First, it put additional moisture into the lamp which was then difficult to remove. It also destroyed the otherwise pleasing appearance of the lamp by causing the silica particles to aggregate and break the coating up with many small cracks. Drying the micronized powder at 775K affected a significant improvement in coating adherence and this became a mainstay of the process for the next several years. This, however, was not the final answer as there was not at that time an efficient method of drying the powder. The 775K dried powder would quickly re-adsorb moisture and the dried powder would also stick to and clog the feed lines of the silica-coating equipment.

Maturing of the Electostatic Silica-Coating Process

Several accomplishments finally made it practical to use unfired silica. The use of plastic and Teflon feed lines reduced the number of agglomerates that caused pinholes in the coating. More uniform micronizing, the addition of TiO_2 to the powder mix and an additional high voltage application following the last coating step improved the adherence of the powder to the glass wall. Finally, the lehring or baking of the lamp after coating but before sealing it, along with the use of maintenance getters, prevented bulb blackening and premature failure due to chemical attack. While the silica-coating process just described has remained basically the same, the concern over ensuring that sufficient moisture is removed from the powder still remained. Better lehring and gettering methods helped but they did not address the main issue of a process that, by its very nature, added water to the lamp.

Dealing with the Water Problem

Water can attach itself to the silica either by filling pores or by chemically attaching to hydroxyl groups on the surface of the silica particles as seen in Figure 1-6.[27] Water can also attach itself to the monolayer of water already attached to the hydroxyl groups. This is called clustering and, along with the primary bonding of water molecules to the hydroxyl groups, is thought to be the method of water adsorption for hydrophilic (water-attracting) silica. When hydrophilic silica is dried, the first water to be removed would be that filling voids or clusters on the surface. When the temperature is raised, water molecules directly attached to hydroxyl groups are released. As the temperature is raised further (925K or more), the surface of the silica changes as the hydroxyls themselves are lost.[27] The loss of surface hydroxyls reduces the capacity for water, but at the same time adversely effects the electrostatic adherence of the particles to the glass bulb.

During the mid-1970s, efforts were again renewed to find a simple low cost coating that would provide similar light diffusing as hydrophilic silica but not the inherent moisture problem. This included non silica powders such as alumina, a reinvestigation into baking the powder prior to electrostatic coating, and the use of hydrophobic silicas.[28] At Westinghouse, this effort was lead by Doug Mattox and Chet Dawson.

Hydrophobic (water-repelling) silicas are formed by the reaction of halogen alkyl silanes with the surface of silica particles. As shown in Figure 1-7, the hydroxyl groups are replaced by organic radicals giving the powder its water phobia. Silica retains its hydrophobic property to temperatures slightly above 775K. Since these temperatures are not encountered in lamp manufacturing, except in the sealing area, this powder was examined as a possible replacement for the hydrophilic variety.

When the prebaking of silica powder prior to lamp coating was first investigated, there was not a practical means to dry it or to keep it dry. This was not the case in the mid-1970s. Lamps coated with prebaked, very low-moisture content (less than 4% loss on ignition) hydrophilic silica performed well even in the hottest of test fixtures; the problems of the powder sticking to and clogging feed lines still existed, however.[29] Lamps coated with hydrophobic silicas also performed well in hot fixture tests but this time adherence particularly in the neck seal area where the powder would blow off during the exhaust process became a concern. A mixture of very finely divided hydrophilic and hydrophobic powders finally solved both problems.[29]

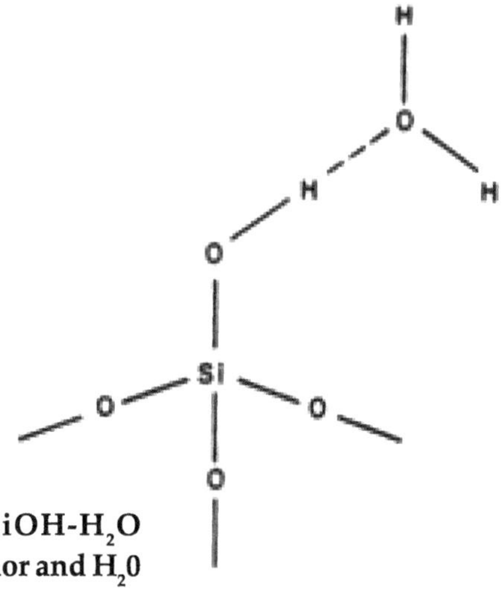

Figure 1-6. Structure of SiOH-H_2O complex, where SiOH is a donor and H_20 an acceptor.

Figure 1-7. One method of forming hydrophobic silica is the replacement of hydroxyl groups with organic radicals.

Cheaper, Better, More Reliable

The electric lamp was probably the best received product of all time. The automobile was initially a product of preference, but light was an imperative. Because the incandescent lamp was so superior to its alternatives, its evolution might have been expected to be deliberate because of its overwhelming customer satisfaction. There was, however, a great sense of urgency from the beginning to make this product better. Perhaps, because its inventors were users as well, they could not resist perfecting the product. Thus, the industry has always pressed toward goals of cheaper, better and more reliable. Early on, these goals translated into the desire for the lamp to be brighter and longer lasting. Major improvements, however, awaited the perfection of the tungsten filament and radically different lamp concepts.

How to Minimize Early "Blow-Outs"

Previously, in "Mechanisms of Lamp Failure," the effects of hot spots on lamp life were discussed. Engineers wanting to improve the reliability of incandescent lamps looked at this problem and first attacked the tungsten wire variability. Certainly, there would always be a thinnest spot, but gross narrowing flaws could be minimized by good filament manufacturing control and this was accomplished; but, what could be done about the over-driving at start up? This overdriving or temperature overshot at a hot spot is a principal reason that many lamps fail when they are switched on. Certainly, regulated power supplies would make lamp installations expensive.

One device, that has been widely marketed through mass distribution publications, was the "bulb miser." A quarter-sized disk was dropped into the socket and the bulb screwed in on top of it. It was made of electroded, multivalent MnO-ceramic aimed at taking advantage of the negative electrical resistance coefficient of ceramics which caused a slow ramp up of the current in the lamp as the ceramic warmed up. In fact, the near doubling of life that resulted was primarily due to an approximate 5V drop across the device leading to the lowering of the voltage and filament temperature.

Development of a Current Limiter

Westinghouse engineers were looking at more effective current-limiting devices in opportunities presented by the emergence of electrically active ceramic materials. During the '50s, Westinghouse scientists and others began to look at a series of ceramic materials which were not the traditional insulators, but were non-linear electrical conductors. These became known by a variety of names depending on their performance, negative temperature coefficients (NTCs), positive temperature coefficients (PTCs), and thermistors (voltage sensitive conductors). Some found their way into electrical motor fuses to kill the power if overheating was detected (PTC) and some became voltage-sensitive shunts (thermistors) which are common today in surge protectors and lighting arresters. One non-linear NTC semiconducting oxide, V_2O_y (y = 3,5) caught the eye of the Westinghouse researchers.

Beginning in 1971, Westinghouse R&D Scientist, George Carlson, guided an effort to develop the first prototype pellets exploiting the NTC characteristics of V_2O_y. The product was made in the form of one-gram pellets by mixing V_2O_3 and V_2O_5 with a phosphate-based glass binder. The milled and blended powders were spray dried and pressed into pellets and fired in nonoxidizing atmospheres. They were subsequently electroded with silver paste and had copper leads attached. The devices were to be inserted in the base of the incandescent lamp.

The device could be made to work in two ways, one of which produced a novel effect. Ordinarily, the device would imperceptibly limit the in-rush current and inhibit early failures. If a more massive pellet were incorporated, the limitation became perceptible and the lamp would come up to full brightness in a gentle crescendo. This was thought to have some attractiveness for formal dining room lighting (reminding the user of a candle's "start-up") or as a bedroom or hall light to minimize the visual shock of turning on the light on arising from sleep in a darkened house

The development was scaled up from research and piloted by Chet Dawson at the development operation in Bloomfield, NJ. From these efforts, it was estimated that the devices would add considerable cost to the lamps. Ultimately, it was decided that there would not be a market for this effective device.

More Efficient Incandescent Lamps

All things radiate a distribution of energies according to their temperatures. If an object is hot enough, it radiates visible light—e.g., the tungsten filament or the sun. Cooler things radiate invisible "light," often infrared energy (heat). Figure 1-8 shows the distribution of light according to temperature for a black body—i.e., non-selective, thermal radiator. The shaded region represents visible light.

We can see how providentially the sun radiates its peak energy in the visible spectrum. If tungsten were operated at its melting point (3600K), its peak radiation would be in the white/yellow part of the visible spectrum. Unfortunately, it would melt quickly at this temperature. At tungsten's normal lamp operating temperature of 2600K, the

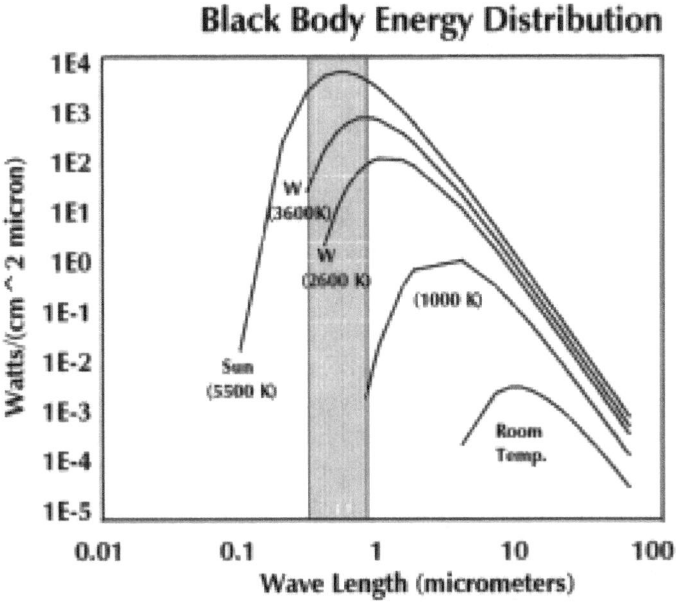

Figure 1-8. The distribution of energy in a blackbody radiator at different temperatures. The two temperatures for tungsten indicate its melting temperature (3700K) and its typical operating temperature (2600K).

blackbody spectrum peaks on the red side of the visible range. Note how room temperature objects, such as furniture, peak in the long infrared. The problem with any black body radiator (temperature induced) is that the greatest fraction of its radiation lies outside of the visible. (This is where discharge lamps can be tailored to have little emission outside of the visible light spectrum.) There is a great deal of wasted thermal energy in incandescent lamps as is easily felt when sitting near them. If the thermal energy could be suppressed, a more efficient lamp would result. This is possible if other than black bodies are used.

A black body is an artificial body which shows no preference for radiating its energy in any particular part of the spectrum but in a smooth, temperature dependent, peaking behavior. The temperature dependence is explained by the Stephan Boltzmann formula:

$$W(\lambda) = \frac{37{,}418}{\left(e^{\frac{14{,}388}{\lambda T}} - \ell \right)} \lambda^{-1} \left(\frac{Watts}{cm^2 \mu m} \right)$$

... where W is the power in watts/($cm^2\mu m$) and λ is the wavelength in micrometers. Two parameters are closely related to the concepts which define a black body, emissivity and absorptance. Emissivity compares a given body to the radiation expected from a black body at the same temperature. If it radiates energy comparable to a black body's, it has an emissivity of 1. In actuality, for real materials, the emissivity usually varies with wavelength. Basically, emissivity and absorption take on similar values—that is, the radiation efficiency is approximately equal to the absorption efficiency.

Some materials show marked preferences for radiating energy over certain specific ranges of the spectrum. If a material had an emissivity of 1 in the visible only and zero everywhere else, it could only radiate its energy in the visible. This possibility was a tantalizing long shot for making an increased efficiency incandescent filament. At Westinghouse, long-ranged investigations were conducted for five years, beginning in the late '60s, to see if such a material could be found. The results were encouraging.

Cermet Filaments

Irv Liberman, a physicist at the Westinghouse R&D Center, built a laser-based apparatus for measuring the emissivity of a range of materials over a range of temperatures. His studies, showed that, while no perfect materials existed, there were some with desirable characteristics. For example, aluminum oxide exhibited a trough in emissivity which swept from the near UV through the visible into the mid infrared. This low emissivity had the characteristic of not radiating energy in the infrared region which would be expected to dominate the radiation in a perfect black body. Aluminum oxide solved the problem of not radiating invisible radiation. Unfortunately, it did not radiate in the visible either. Further, aluminum oxide was an electrical insulator and could not be heated electrically.

The heating problem was given to Larry Cadoff in the ceramics group at the Westinghouse R&D Center. Cadoff attempted to produce cermets using mixtures of aluminum oxide and molybdenum. Cermets are ceramic-metal mixtures, where it was hoped that metal continuity would be used for resistive heating. Cadoff mixed the metal and ceramic powders in polymer binder/plasticizer mixtures and extruded filamentary polymer shapes. These were fired in a non-oxidizing atmosphere to remove the polymer and produce rigid straight wire shapes.

The remaining element of a solution was suggested in a collaborative effort with another ceramic engineer in the materials group, Bulent Yoldas. Yoldas had been carrying out pioneering work in polymerizing oxide materials from organic precursors. In particular, this allowed for the possibility of mixing ceramics at the atomic level in their liquid precursor form. Yoldas added minute amounts of chromium oxide to the aluminum oxide, producing the characteristic ruby tint in the ultimately fired aluminum oxide. The Cr^{+2} dopant raised the emissivity in the visible preferentially.

All that remained to accomplish was to combine these ideas into a cermet filament lamp. Some prototype filaments were self-heated to incandescence and operated in evacuated laboratory envelopes. Figure 1-9 shows a laboratory lamp made from this shaped or tailored emitter. Eventually, the commitment to the project gave way to the severe business curtailment experienced by the lamp industry in 1973. The lamp industry was asked to bear a disproportionate share of the burden of the national energy conservation program which had begun following the middle east oil import embargo. The federal campaign, which asked the commercial sector to idle every other lamp, sent a crushing blow to sales and strategic investment throughout the lamp industries.

Alternative High Efficiency Incandescent Strategies

The shaped emitter filament was a bold concept which could only be approached by a deliberate long-range research effort. The oil embargo of 1973, however, rallied the nation around quick fixes. Some timely progress in solar collector research offered some promise for enhancing the efficiency of the incandescent lamp through a different strategy. Conceding that the bulk of tungsten radiation is in the form of heat (infrared energy), could a coating be found which would, in essence, be a heat mirror? Specular metals could fill that role, but they were opaque to visible light. Might there be a visibly transmitting IR reflector, which would reflect the thermal energy back to the filament where it could be used to maintain the temperature of the filament? This would produce a down conversion of radiant energy, increasing the energy in the visible portion of the spectrum, using less electrical power.

MIT researchers, Fann and Bachner, discovered a particularly effective three-layer optical film which had exactly those properties, and which they patented for application to solar collectors.[30] Each layer was extremely thin—180 angstroms. Layers of TiO_2-Ag- TiO_2 were applied to

Figure 1-9. A laboratory incandescent lamp with selective emitter prepared from a chromium-doped, aluminum oxide molybdenum cermet. Courtesy of Philips Lighting.

the large flat glass panels of the collectors by sputtering. Little attention had been paid to their efforts by the lamp community until a specialty lamp manufacturer in New Jersey, DuroTest Corp., acquired an exclusive license of the MIT patented technology for application to incandescent lamp manufacture.

A substantial publicity campaign was initiated by DuroTest to capitalize on its momentum. The early publicity reported reliably that DuroTest had contracted with Materials Research Corporation of Orangeburg, NY, to develop specialty sputtering equipment for applying the three layer coatings to the interior of incandescent lamps. Reac-

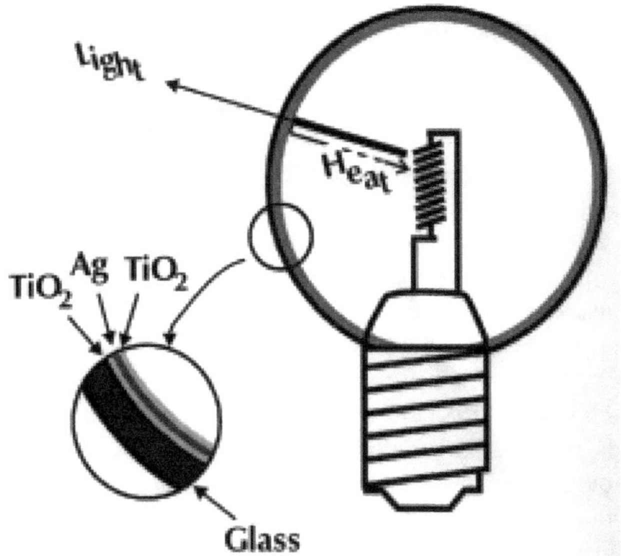

Figure 1-10. The design of the high-efficiency IR-reflecting visibly transmitting coated incandescent lamp. Three layers of TiO_2-Ag-TiO_2 coating, 180 angstroms each, are applied to the interior of the lamp.

tive sputtering was to be used for the TiO_2 layers and simple sputtering would produce the silver layer. The concept would resemble the schematic drawing in Figure 1-10. As Westinghouse responded to this lamp concept, it recognized that the economics of three-stage internal sputtering would be formidable and that internal sputtering would be problematic. Fortuitously, Westinghouse R&D Scientists had been developing low-cost optical coating techniques using organic-based oxide solutions.

Bulent Yoldas, who was mentioned in connection with the shaped emitter, had considerable success synthesizing a number of ceramic oxides from sol-gel precursors. He was a prolific patenter and publisher. His early publications attracted a lot of imitators and competitors, but the publications were on average about 3-4 years behind the laboratory and patent technology. On the promise of his early publications concerned with producing monolithic ceramics by sol-gel techniques, imitators and rivals rushed forward to produce bulk ceramics and high-purity, fine-particle powders, but at the R&D Center, Westinghouse had heavily pressed this technology in a totally different direction—thin films.

Yoldas saw how effective these solutions were for coating metals and ceramics with nanometer-thick dense ceramic layers. He was invited to attempt the development of an anti reflective coating layer for dendritic web drawn silicon for photo voltaic applications. This proved to be extraordinarily successful, resulting in the application of optically controlled, thin films of TiO_2 by a simple continuous dipping and baking process which was quickly patented.

It occurred to all concerned that this solution process ought to be equally successful in coating the interior of an incandescent lamp, and at low cost. Westinghouse envisioned producing the Fann and Bachner type films by the very economical process of fill-drain-bake TiO_2, sputter Ag, fill-drain-bake (725K) TiO_2. The company quickly obtained the prototypical (4 in. spherical) bulbs from Corning Glass Works and tested this technology. The results were very successful and a number of prototypes were developed for transporting to the Materials Research Corporation in Orangeburg, NY, for controlled sputtering. The Materials Research Corporation was contracted to deposit controlled silver layers on pre-coated TiO_2 bulbs. The bulbs were returned to the Westinghouse R&D Lab for a final TiO_2 coat. The results were excellent and ultimately, a patent was applied for and issued.[31]

The Arrival of the Compact Fluorescent Lamp

The timing again proved to be very poor, since the energy crisis began to ease and more promising compact fluorescent lamps began to be attractive. Strategic decisions at Westinghouse Lamp Headquarters took a position of leaving these options in reserve while concentrating resources on the compact fluorescent lamp. This proved to be an economically beneficial decision as efforts to develop the efficient incandescent lamp with heat mirror revealed that the costly perfection of the envelope shape was a very important factor in overall lamp efficiency. The refocused reflected thermal energy was key to success and this required an excessive premium on the envelope and filament centering. Westinghouse had proved to have developed a workable technology, but there were alternative solutions of greater promise—i.e., the compact fluorescent. Under the most favorable conditions, theory and experiments showed a maximum efficacy of about 30 lumens per watt could be achieved.[32] This is considerably less than the compact fluorescent would provide.

The Future of The Incandescent Lamp

While office, factory, street and recreational lighting have almost entirely switched to more efficient lighting sources, the incandescent lamp continues to be the choice for most residential and decorative lighting uses plus many specialty applications. Compact fluorescent—and to a lesser extent, halogen lamps—have begun to fill a small share of general service incandescent sockets. Their higher cost and a general customer preference for the convenient and familiar incandescent lamp, however, have slowed their wider acceptance.

There will never be another surge in incandescent lamp research as occurred in the '60s and '70s. Today, the R&D expenditure is almost exclusively application- or cost reduction-oriented. New products, however, continue to appear. For example, modern electronics have made it possible to place an application-specific integrated circuit, in the form of a small button shaped disc, into the base of incandescent lamps. These devices can cause the lamp to turn off automatically after a preset time or dim to provide a night light. Another example is the use of computer-aided design leading to improved and more efficient shapes for reflector lamps. Cost and environmental concerns have directed efforts to eliminate lead and cadmium from all lamps.

The future for incandescent lamps is expected to remain as it has been for the past decade. A slow decline in the use of general service lamps will be offset somewhat by a continued strong demand for decorative and specialty lamps. Without a driving force for research expenditures, the knowledge base for incandescent lamps will continue to slow. Incandescent light sources will still make our world brighter in the 21st Century, but they will never again have the prominence they once had.

Acknowledgements

The authors wish to thank Dr. M. Mujahid, Dr. Jack Singleton and Wanda Calvin for their assistance in gathering information and providing valuable input. Also our wives, Marie Mattox and Pat Dawson, for their invaluable support, assistance in searching the Web and editing.

References

1 James A. Cox. *A Century of Light.* New York: The Benjamin Company, Inc., 1979.
2 John W. Howell and Henry Schroeder, "History of the Incandescent Lamp," The Maqua Company, Schenectady, NY, 1927.
3 Private Discussion with Dr. Jack Singleton, June 1995.
4 D.P. Smith. *Hydrogen in Metals.* Chicago: University of Chicago Press, 1948.
5 E.A. Gulbransen & K.F. Andrews, "Mechanism of the Reaction of Hydrogen with Zirconium," *Journal of the Electrochemical Society,* Vol. 101, No 7, July 1954.
6 Dr. E. Robusin, "Gettering for Lamps," SAES Getters, S.P.A., Milan, Italy, Oct. 1978.
7 J.B. Murdoch. *Illumination Engineering: From Edison's Lamp to the Laser.* New York: Visions Communications, 1994.
8 *Westinghouse Lighting Handbook,* 1980.
9 R. Becker, "The Life Period and the Vaporization of Tungsten," Zeitschrift f,r Technische Physic, Vol. 6, 1925, p.309.
10 H. Horster, E. Kauer & W. Lechner, "A Concept for the Burning Out Mechanism of an Incandescent Tungsten Wire," *Journal of the IES,* July 1972.
11 F. Harvey, "Failure of Incandescent Tungsten Filaments by Hot Spot Growth," *Journal of the IES,* July 1974.
12 G. Comentz, H.A. Johansen, J.W. Salatka, "Hot Spots Caused by Necks in a Straight Incandescent Lamp Filament in Gas," *Journal of the IES,* July 1973.
13 C. Dawson, "Effect of a Temperature Gradient on Bubble Growth in Tungsten," *Metallurgical Transactions,* Volume 3, December 1972.
14 C. Dawson, J. Nelson, W. Radzelovage and H. Sell, "Hot Spots and Their Effect on the Lives of Incandescent Filaments," *Journal of the IES,* July 1973.
15 C.J. Phillips, "Glass the Miracle Maker," Pittman Pub. Co., 1941.
16 T.A. Edison, U.S. Patent 223,898, "Electric-Lamp," Jan. 27, 1880.
17 E.B. Shand. *Glass Engineering Handbook,* Second Edition. New York: McGraw-Hill Book Co., Inc., 1958, pg. 6.
18 A.K. Varshneya. *Fundamentals of Inorganic Glasses.* San Diego: Academic Press, Inc., 1994, pg. 335.

19 A.A. Bright, Jr. *The Electric lamp Industry*. New York: The Macmillan Company, 1949, p. 326.

20 F.A. Lewis. *The Incandescent Light*. New York: Shorewood Publishers, Inc., 1949, p. 105.

21 P.W. Keating. *Lamps for a Brighter America*. New York: McGraw-Hill Book Company, Inc., 1954, pp. 126-128.

22 M. Pipkin, *General Electric Review* (March 1949), pp. 14-17.

23 M. Pipkin, US Patent 2,545,896 granted to the General Electric Company, (March 20 1951).

24 M. Pipkin, US Patent 2,626,874 granted to the General Electric Company, (Jan. 27, 1953).

25 G. Meister & F. Cerulli, US Patent 2,922,065, granted to the Westinghouse Electric Company (Jan 19,1950).

26 G. Meister & G. Young, "Electrostatic Coating of Incandescent Lamp Bulbs" A Westinghouse Internal Report (Jan. 9, 1950).

27 K. Killer, J.H. Shen & A.C. Zettlemoyer, "Water on Silica Surfaces. I. Partially Hydrophobic Silicas," *The Journal of Physical Chemistry*, Vol. 77, No. 11, 1973.

28 D.C. Henderson, C. Falls & K.M. Maloney, " US Patent 3,868,266 granted to the General Electric Company, (Feb. 25, 1975).

29 C.W. Dawson, R.B. Parsons & D.M. Mattox, " US Patent 4,099,080 granted to Westinghouse Electric Corp. (July 4, 1978).

30 J.C.C. Fan, F.J. Bachner, G.H. Foley, Pl M. Zavracky, "Transparent Heat-Mirror Films of TiO2/Ag/TiO2 for Solar Energy Collection and Radiation Insulation," Lincoln Laboratory Report, Industrial Liaison Program, 11-42-74, 1974.

31 B.E. Yoldas, U.S. Patent 4,293,593, "Method of Fabricating Heat Mirror for Incandescent Lamp Envelope," Oct. 6, 1981.

32 *Energy Efficient Incandescent Lamp*, Final Report, DOE-LBL Subcontract 4 506 010, Dec. 16 1980 - Nov. 30, 1981, LBL-14546, EEB-L-82-06, L65.

Chapter 2

The Potassium Secret Behind Modern Tungsten Wire Production

by Alfred G. Gahn, Georg E. Marsen,
David M. Moon and Heinz G. Sell

Introduction

By the early 1900s, the carbon-filament lamp had reached its peak of luminous efficicacy with 3.5 to 4.3 lumens per watt (LPW). The quest for improved filament material continued and among others, tungsten was a leading candidate. This was obviously because of tungsten's high melting point of approximately 3400°C, the highest of all known metals, and consequently its rather low vapor pressure at operation temperatures of incandescent lamp filaments. While platinum, osmium and tantalum showed a significant improvement over carbon-filament lamps, the life of these lamps was poor due to recrystallization effects on alternating current, which was the dominant electrical mode already at that time. The first lamps with tungsten wire achieved a sensational efficacy of 7-8 LPW at a user life of 800 hours.[1]

Production of wires for filaments was nearly impossible, however; tungsten was found intractable, either as a brittle mass or as a fine crystalline powder, that resisted being worked into a coherent form. These annoying metallurgical properties were certainly the reason to place tungsten into the category of *refractory metals*, a family of metals being "obstinately resistant to manufacturing or to controlling."[2] How to convert this stubborn material into filament wire for a U.S. market that had grown to more than 50 million sockets for incandescent lamps was a question being actively researched.

During the 17th assembly of the American Electro-Chemical Society at Pittsburgh in May 1900, C.G. Fink of the General Electric Company (GE) in Schenactady, NY reported the first success of a stretchable tungsten wire achieved by hot working methods.[3] Subsequently, William D. Coolidge, also at GE, developed a process which converted tungsten powder into a "ductile" wire that could be wound into coiled filaments.[4] Coolidge's work with GE began in September 1905 (he worked on tungsten until 1945), and by 1910 GE announced Coolidge's first success in producing ductile tungsten wire. The process consisted of the following major steps:[4]

1. Pressing tungsten-powder into a bar of 6.5-9.5mm square and 150mm long.
2. Sintering the bar at high temperatures in hydrogen into an ingot.
3. Hot hammering (swaging) the ingot at high speed rotation (several 100 rpm) using two die-halves. Successive smaller dies led to a round wire of 0.76mm in diameter.
4. Drawing the wire through heated diamond dies. Staggered reduction steps of some 20μm yielded a tungsten wire of 25μm(1/1000 inch).

Although this tungsten wire was fairly ductile, it still had some unpleasant limitations concerning lamp operation. Filaments wound from this wire sagged during lamp operation and, when touching the outer glass envelope, melted a hole into it, resulting in premature lamp failure.

This problem was solved after much experimentation by Aldar Pacz,[5] who was also working for GE in 1917. The tungsten powder so far had been produced from "high purity" WO_3 (yellow tungsten oxide) by firing it in ceramic-based dishes (so-called Battersea Dishes, from where it is likely the German abbreviation BSD for doped tungsten derived). This tungsten powder contained, among other impurities, measurable amounts of potassium extracted from the crucibles' material.

Since wire manufactured from this powder exhibited less sagging than pure tungsten wire, A. Pacz reproduced the effect of potassium by deliberately doping the tungsten oxide with oxides of potassium, silicon and aluminum (AKS-doped tungsten). Empirically, he optimized the properties of tungsten wire, and could finally avoid the sagging with a batch, "trial No. 218", which led to the code name "GE 218," which is still used today. It was the essential processing step, doping, which led to the required properties of AKS-doped tungsten, and thus the door was opened to mass supply a satisfactory tungsten wire to a fast-growing lamp industry. It heralded the demise of the traditional carbon filament lamp for general lighting purposes.

Table 2-1 provides the U.S. production volumes of General Lighting Service (GLS) Lamps between 1885 and 1945. From 1919 onward, the lamp exclusively used was the tungsten filament lamp. The miniature lamp market also rapidly developed, from 85 million lamps in 1921 to a staggering volume of 1.1 billion lamps in 1945. A similar fast market growth happened in Europe.

Table 2-1. General lighting service market development (large lamps) in the United States between 1885 and 1945.

Year	*Annual Production (Millions)*	
1885	0.3	100 % Carbon Filament
1891	7.5	100 % Carbon Filament
1909	66.7	80 % Carbon Filament
1914	88.5	15 % Carbon Filament
1919	224	Nil Carbon Filament
1925	273	
1929	343	
1932	334	
1935	413	
1940	593	
1945	795	

On-going product and process development work with a tremendous effort in machine design and construction was required to realize these production figures.

Modern Tungsten Powder Production

Research Progress: 1950-1996

While the unique properties of AKS-doped tungsten propelled incandescent lamps into prominence, the reasons as to how they came about, often referred to as the doping mechanism, remained a mystery until the second half of the twentieth century.

After World War II, all incandescent lamp markets grew rapidly, which put an equivalent demand on raising filament grade wire production. Obsolete tungsten powder and tungsten wire manufacturing facilities limited manufacturing capacity, which in turn led to serious quality problems, notably breaks in wire drawing and splitting in filament winding. This situation forced the major lamp manufacturers to devote increased research and engineering efforts into solving these urgent material problems and gaining manufacturing capacity with state-of-the-art facilities. It was therefore, only natural that, as part of these improvement programs, research was directed toward learning what role the AKS dopants played in controlling the metallurgical properties of doped tungsten. This research bore fruit with the introduction of high power electron microscopy and modern analytical and diagnostic methods.

In addition, in its chemistry, tungsten supplied reasons to be called "refractory." Hardly any results on tungsten oxides from early diffraction experiments proved valid under the eye of electron small area diffraction. A new approach to stoichiometry had to be developed in order to understand the nature of this apparently simple compound. "Shear phase" and "multivalence" were concepts which had to be applied in order to make the results of X-ray investigations compatible with those of elementary analysis. But the strange influence of what was, at that time, considered minor traces of potassium in the metallurgy of the highest melting metal was still a secret. The first steps to a better understanding were made when small angle X-ray diffraction proved that a second phase exists in doped tungsten wire.

What that second phase was only became apparent after another 20 years of instrumental progress and, thinking the unthinkable, one had to admit that potassium metal can coexist with tungsten during sintering and swaging at temperatures as high as 15 times its melting point. This unique discovery challenged the scientists to solve this puzzle. In the end, metallurgical and chemical investigations opened the way to a complete understanding of this extraordinary symbiosis.

The Chemistry of Ammoniumparatungstate (APT)

Ammoniumparatungstate (APT), as the starting material for all tungsten going into lighting applications, is one of the crucial chemicals where even diminishing traces of other elements can drastically change the performance of a lamp wire. So, for many decades, all major lamp manufacturers started with ores such as Scheelite, Wolframite and Huebnerite or with tungsten trioxide and followed a well established hydrometallurgical process of repeated dissolution and precipitation in water to remove all traces of deleterious impurities, in order to produce pure APT. Nowadays, APT is commercially available with satisfactory quality from China and other countries.

APT is a heteropoly salt of tungsten with two non-acidic hydrogen ions as central atoms. The structure is referred to as Keggin polyhedron (see Figure 2-1) and is best described by the formula:

$$(NH_4)_{10}\,[H_2W_{12}O_{42}] \times 4H_2O$$

... where the amount of crystal water may vary within certain limits. As a given and well-known compound, chemists in the lamp industry took little interest in the understanding of that compound besides how to produce it with the desired quality. Their main interest was directed to how to decompose it.

Doped Tungsten Powder Production

The production of doped tungsten metal powder ready for pressing and sintering into ingots involves five basic process steps (see Figure 2-2):

Figure 2-1. Polyhedron model of the paratungstate anion $[H_2W_{12}O_{42}]$ with two centered non-acidic protons.

First Reduction

Reduction of APT to a tungsten oxide of "blue" color. The reduction can be carried out in a rotary furnace or in a pusher furnace loaded with boats made of a special alloy, such as Inconel, under reducing atmosphere and at temperatures of between 450°C and 850°C.

Doping

The "blue" tungsten oxide is slurried with a water solution of $K_2Si_4O_9$ (KASIL), adding potassium chloride and aluminum chloride or nitrate followed by air drying of the mixture.

Second Reduction

The dried, doped "blue" tungsten oxide mixture is reduced in hydrogen of defined dew point in pusher furnaces loaded in metal boats. Particle size control is achieved by controlling the dew point of the hydrogen and by the filling height of the boats for a given temperature/time profile of the furnace tubes.

Washing

The as-reduced doped metal powders are washed in diluted hydrofluoric acid to remove deleterious impurities such as Fe and Ni, followed by air drying and sifting for particle size range.

Blending

Different particle size fractions of the washed doped metal powder are blended into a compressible and workable "master batch" of an ideal particle size distribution. Normal particle size ranges from 3 to 5 microns, as usually determined by the Fisher Sub Sieve Size (SSS) test. A scanning micrograph representative of doped tungsten powder is shown in Figure 2-

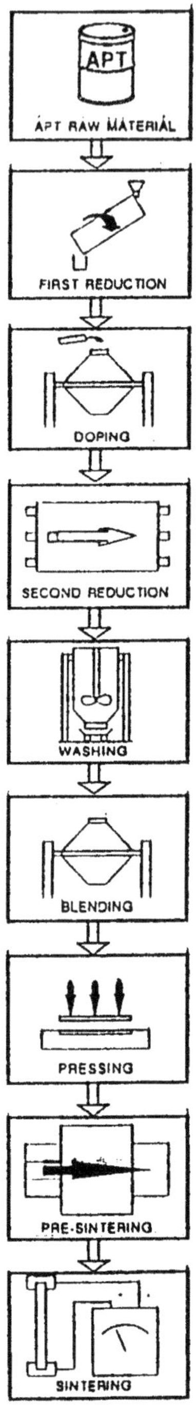

Figure 2-2. Schematic of the production of doped tungsten ingots.

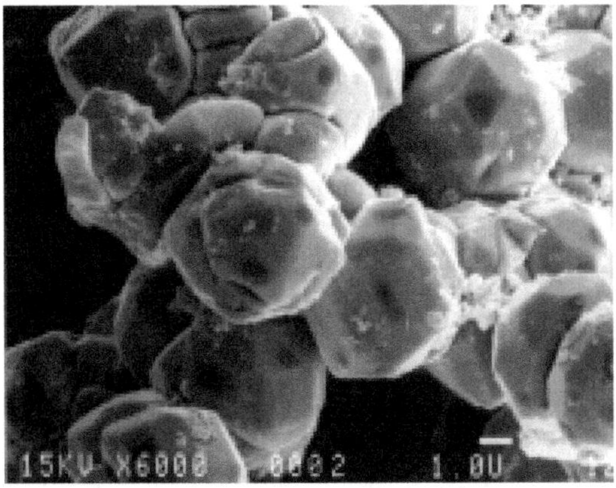

Figure 2-3. Scanning electron micrograph of AKS-doped tungsten metal powder (6000X) (dark globs on particle surfaces are rich in potsassium).

3.

As points of reference, in elemental form and depending on particle size fraction, doped metal powders retain:

Aluminum (Al)	50-200 ppm
Potassium (K)	100-500 ppm
Silicon (Si)	200-700 ppm

...of the originally added amounts of dopants.

The Funny Story of "Blue Oxide"

When APT is decomposed in air or in another oxidizing atmosphere, it turns into yellow tungsten oxide, WO_3, under evolution of ammonia and water. WO_3 seems to be a well-defined compound, but when adding doping elements in the form of potassium silicate and aluminum salts to this yellow oxide, the outcome is not always the same. WO_3 is a polymorphic oxide, the reactivity of which depends on its history. Consequently the older doping process starting with WO_3 was not reproducible in regard to the composition of the dopant compounds which formed.

Based on experience, most tungsten manufacturers therefore changed their doping process in the '70s to start by slurrying the dopants into the "blue oxide" mentioned before, because it gave them greater reproducibility. This "blue oxide," with the nominal formula $W_{20}O_{56}$, stimulated more research.

However, whatever analytical tool was used, there were hardly ever two lots of "blue oxides" with identical crystallographic phase composition. Finally, another family of tungsten compounds, the hexagonal tungsten bronzes (see Figure 2-4), caught the scientists' attention.

These bronzes, known for about a century, are highly stoichiometric, with tungsten to oxygen ratio 1:3, but containing a remarkable amount of alkali as high as 0.5 mole fraction.[6]

The "blue oxides" used for doping turned out to be a mixture of two lower tungsten oxides: one identified as ammonium-tungsten-bronze $(NH_4)_xWO_3$ and the second an amorphous portion of unknown composition amounting to some 40 atom % of the total tungsten involved.[7] Practical results showed that a high temperature treatment of APT to over 850°C leads to a better-defined product which is less prone to variations within the doping process.

The Chemistry of Doping

This section deals with the kinetic effects which made the doping of "blue oxide" so successful compared to the same process step with "Yellow Oxide" (WO_3).

Using salts of potassium, silicon and aluminum for doping either with WO_3 or with blue oxide, it was noted that in both cases the same chemical reactions occur. Besides the expected alumosilicates and silicates, it was discovered that heteropoly salts also play a major role in the process.[8,9,10] Two species, the potassium-undecatungstenosilicate K_8 (Si W_{11} O_{39}) and the potassium-dodecatungstenosilicate K_4(Si W_{12} O_{40}) were found by UV-spectroscopy to co-exist in a pH-sensitive equilibrium. Subsequently, it was shown that with a "blue oxide" containing some 30-40% of ammonium tungsten bronze, the formation speed of

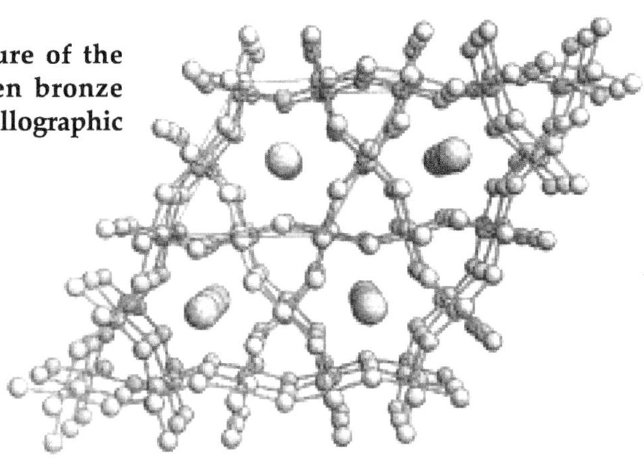

Figure 2-4. Structure of the hexagonal tungsten bronze seen from the crystallographic C-axis.

these compounds was much faster than in a yellow oxide. Actually, none of the two oxides come to an equilibrium within the technical process times, but the more acidic behavior of the bronze-containing oxides brings it closer to the ratio:

$$\frac{c\left(Si\ W_{11}O_{39}\right)^{8+}}{c\left(Si\ W_{12}O_{40}\right)^{4+}} = 0.8$$

Scanning Auger microscopy revealed that with an equilibrium ratio of 0.6-0.8, these compounds form a low melting phase, which provides high ionic mobility for potassium at the early stages of the reduction to tungsten metal.

It is important to know that about 60-70% of K, 12-15 percent of Si but none of the aluminum enters into this reaction. The remaining amounts, according to analytical findings, show surprisingly molar ratios corresponding to a nominal formula of $KAlSi_3O_8$, the well known feldspar. However, a "feldspar" has never been identified by X-ray analyses. This is not surprising considering the aqueous environment during doping. Feldspar is one of the most stable potassium compounds known. It is believed that the excess of amorphous feldspar joins in the salt reaction with the tungstenosilicates.[11]

The Role of Beta-Tungsten

The nature of β-tungsten was, for a long time, only discussed with moderate academic interest, but was brought back into the scientific literature as a possible—although disappointing—candidate for a superconductor because of its A15 structure. But, in contrast to α-tungsten, it has some properties which makes it contribute to an understanding of the doped tungsten wire behavior for incandescent filaments.

Beta-tungsten is capable of dissolving and absorbing large amounts of other elements and ions including potassium, aluminium, silicon and oxygen. These elements are dispersed atomically in the lattice of β-W. During the diffusionless transition into the smaller unit cell of α-tungsten, these ions are trapped. On further heating under hydrogen, reduction and oxidation of the elements take place. The radius of the oxygen atom is smaller than that of the ion and consequently can migrate to the surface, where it is removed by reaction with hydrogen. Potassium atoms, on the other hand, being larger in size by a factor of approximately 1.6, become immobile and remain trapped in the lattice. Newer investigations showed that aluminum stabilizes the β-tungsten phase.[12] As a consequence of the longer life of this phase, more potassium from the reaction can diffuse into the β-tungsten A15-lattice—i.e., the aluminum level (up to 3%) controls the reproducibility of the doping process.

We can conclude, therefore, that silicon is essential for the doping process in that it "mobilizes" potassium in the salt melt at moderate temperatures, whereas aluminum secures the stability of β-tungsten over a wider range at temperatures, where the incorporation of potassium becomes possible. All attempts to directly dope pure α-tungsten with these AKS elements have never been successful.

Sintered Ingots:
The Nursery of Potassium Bubbles

The process steps from tungsten powder to fine wire are outlined in Figure 2-5. This section describes the technology of converting blended doped tungsten powders into bars of rectangular, hexagonal or circular cross sections weighing 1-6 kilograms and being 80-120 cm long. Bars of rectangular or hexagonal cross sections are produced by mechanical die-pressing the powders in appropriately configured steel molds at about 40 tons/square inch. Bars of circular cross section are obtained by hydraulically pressing the metal powders in rubber molds. The as-pressed "green" bars are, however, fragile and therefore delicate to handle. To increase their fracture strength, the "green" bars are presintered under hydrogen at 1200-1400°C for 15-30 minutes. This step not only increases their fracture strength, but also raises their density to 60-70% of tungsten's theoretical density (19.25 g/cm^3).

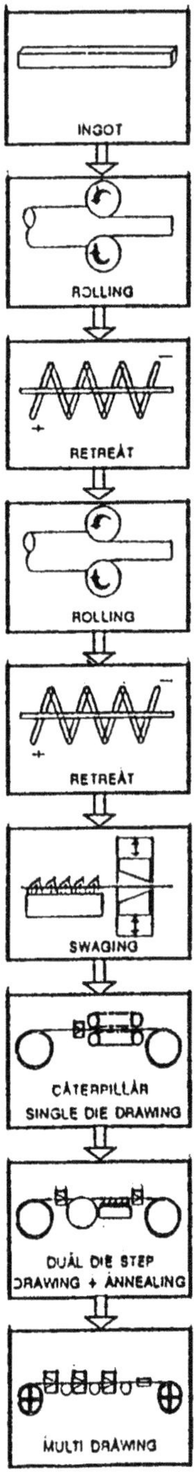

Figure 2-5. Schematic of the production of doped tungsten wire for incandescent filaments.

Following the presintering step, the bars are mounted into sintering bottles—one per bottle—where they are heated in hydrogen to temperatures as high as 2900°C. A view of a typical production set of sintering bottles is shown in Figure 2-6.

To sinter, the bars are heated by passing AC current through them according to an empirically determined time/current schedule (see Figure 2-7).

During sintering, most of the dopants evaporate from the ingots. To control the outgassing, the time/current schedule includes two constant temperature periods—one at 2100-2200°C, the second, the so-called high point, at 2800-2900°C . Due to the declining resistivity, when the ingots densify in the course of sintering, the practice is to control the current by voltage adjustment.

The first holding period at a temperature of 2100-2200°C creates an "open porosity" in the ingot, which allows most of the dopants to escape into the surrounding flush gas. Newer advanced analytical methods have shown that during this holding period Al, Si and oxygen are eliminated close to the low final content in the ingot, while K is retained at approximately two-thirds of its level in the tungsten powder (see

Figure 2-6. Bottle sintering facility for tungsten ingots.

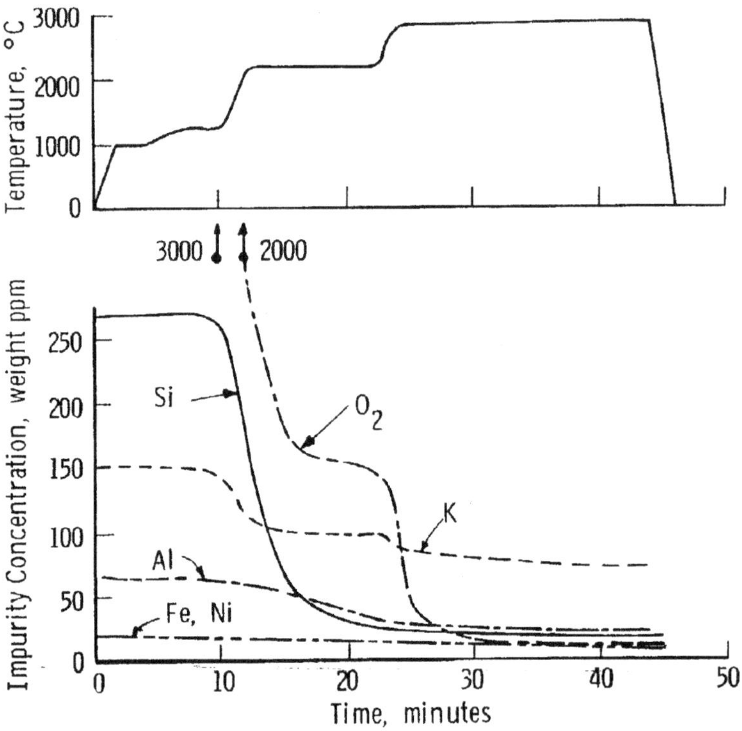

Figure 2-7. Typical sintering schedule for AKS-doped tungsten and impurity content at various stages of sintering.

Figure 2-7).

On the rise to the second holding period at 2700-2900°C, the pores close and most of them shrink drastically, even disappear by vacancy diffusion. It is during this phase that potassium is trapped and potassium-containing pores are created under relaxation of the tungsten lattice. These pores reach their equilibrium size, governed by the surface tension of tungsten and the internal potassium vapor pressure, at the sintering temperature.[13,14]

That these pores in fully sintered ingots are, indeed, formed by metallic potassium was confirmed by Auger electron spectroscopy, secondary-ion mass-spectrometry and dispersive X-ray analysis.[15] The investigation was performed on AKS-doped tungsten as well as on similarly treated pure tungsten and on electron-beam melted tungsten. The materials differed very significantly in their bulk potassium content. A typical potassium containing pore distribution of a sintered ingot is summarized in Table 2-2.

Table 2-2. Bubble size distribution (in percent) in sintered and rolled ingot.

Bubble Size (mm)	*Sintered Ingot*	*Rolled Ingot*
0-0,5	1%	2%
0,5-1,0	50%	51%
1,0-1,5	32%	24%
1,5-2,0	11%	9%
2,0-2,5	4%	5%
>2,5	2%	9%

Processing Ingots into Tungsten Wire

Rolling And Swaging

Following the teachings of the Coolidge process, production consists of approximately 60-70 deformation steps to yield tungsten wire of 18μm from sintered ingots. In addition to the deformation steps, the process includes retreat (recrystallization), anneal (softening) and surface-cleaning steps. The process differs somewhat from manufacturer to manufacturer depending on ingot size (weight) and cross-section.

Deformation processing begins with breakdown working of the sintered ingots. For that purpose, the ingots are preheated to 1400-1500°C for several minutes to be uniformly heated from end-to-end. In the particular process depicted in Figure 2-5, they are then passed through a tandem mill (see Figure 2-8), undergoing 6-8 reductions in cross-section.

After passing through the mill, the ingots have been deformed into long rods, which are ejected from the mill glowing red hot and steaming tungsten oxide that evaporates from the surface. The rolled rods are cooled down in air or submerged in water, to which sometimes etching agents are added to clean the surface from impurities, such as iron, picked up from the rolls of the mill.

In earlier times, ingot breakdown working was predominately carried out by swaging instead of rolling. Swaging involves hammers which rotate at high speed. The same as in rolling, the ingots are first preheated, however, because of the much slower passage of the ingot through the swaging machine (3 meters/minute), temperatures as high as 1800°C are used before they are fed into the swager. Breakdown swaging is also carried out in air. As the tungsten has reached nearly

Figure 2-8. Six-stand tandem mill used for breakdown rolling of sintered tungsten ingots. F = feed mechanism, C = RF preheating system, S = six-stand mill.

theoretical density at that stage, only a thin layer of WO_3 forms on the surface of the rod, while the bulk of the metal is not affected by the oxidation. Swaging is generally still used for rods between about 7 and 4mm in diameter.

Drawing Techniques

At about 4mm diameter, the ingots have been elongated into rods of unmanageable length. Meanwhile, the tungsten grains of the ingot have been transformed into long fibers, which makes drawing through carbide dies followed by winding of the wire onto larger size capstans possible. This is a crucial step in that one has to avoid inhomogeneities of the microstructure which can cause splits and cracks. Large deformation at elevated drawing temperatures is required to bring about the defect-free drawing texture required to avoid wire breaks in wire drawing.

In this manufacturing process (see Figure 2-5) for single-step drawing between 4 and 1mm diameter, the heavy deformation followed by spooling is accomplished on a caterpillar drawing machine (see Figure 2-9).

To reduce friction in the carbide dies and obtain smooth wire surfaces, the heavy wire is lubricated with finely dispersed graphite mixed with additives according to an empirically developed formula. The graphite provides the "lube effect" when baked onto the wire surface. It protects the rod and wire from excessive oxidation at the drawing temperature of 1000-1200°C. This drawing sequence has the effect that the

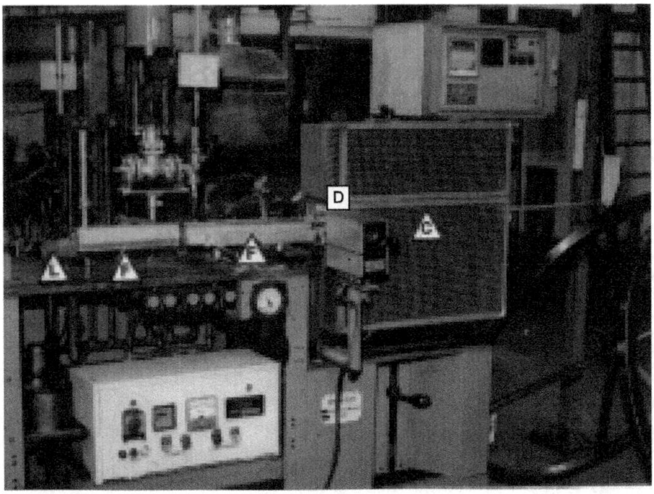

Figure 2-9. Caterpillar heavy wire drawing machine. L = lubrication, F = gas-fired lube drying and wire preheating, D = die system with IR temperature control, C = caterpillar draw.

fibers have become much smaller in cross-section and much longer, a structural condition which greatly facilitates defect-free drawing and allows the drawn wire to be spooled on capstans of smaller diameter.

In fact, as a technological advance, wire in the medium size range, from 1mm to 0.5 mm diameter, are now routinely drawn in tandem through several dies simultaneously. A typical multi-die wire drawing machine for this wire size range is shown in Figure 2-10.

As the wire size decreases, the drawing temperature is also reduced. Below 0.5 mm, multihead drawing techniques are used, where the tungsten wire passes simultaneously through 4-8 dies, while at the same time it is being preheated and lubricated. A modern fine wire multidrawing machine (4 reduction steps) for wire below 18μm is shown in Figure 2-11.

The transfer from die to die is effected by a cylindrical or conical capstan which is surface-coated with alumina, zirconia or chromia.

While tungsten carbide/cobalt materials are used for larger diameter drawing dies, wires below 0.5mm are drawn through dies of synthetic polycrystalline diamonds or natural diamonds. Lubrication is always graphite-based, however, declining in concentration with decreasing wire size.

Figure 2-10. Triple die medium-size wire drawing machine. F = gas-fired burners, C = capstan, D = die system, S = split tester, Sp = spooler.

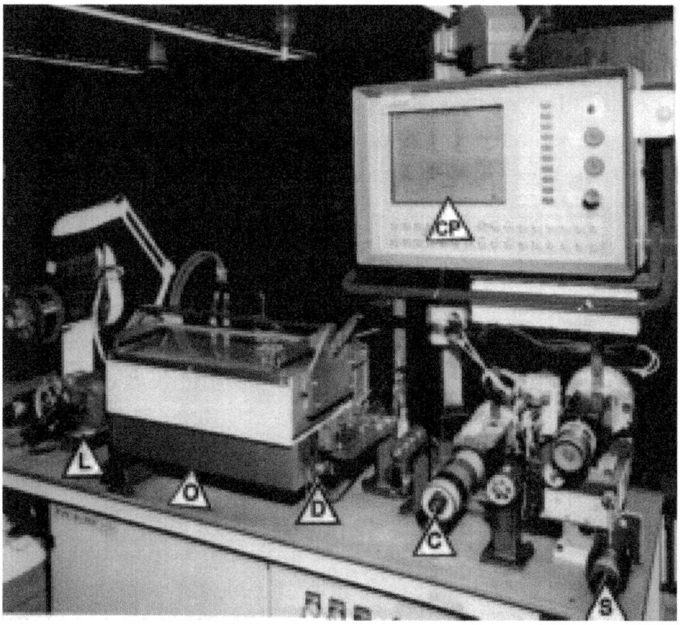

Figure 2-11. Fine wire (1.0-0.7mg/200mm) multi-die drawing machine. L = lubrication, O = electrical oven, D = die system with temperature control, C = capstan, S = spooler, CP = electrical control panel.

The Role Of Graphite Lubricant

Since the invention of Coolidge, some form of colloidal graphite—natural or synthetic—has been used to coat the tungsten wire prior to the drawing steps. Pure experience led to the practical recipes used by the wire manufacturers. Unfortunately, only little research was carried out to explore how the graphite and the additives work. Some research on lubricants and their use has recently been published by M. Lee and D.G. Flom[16] and by W. Martens and P. Brulez.[17] There is no doubt that a tungsten oxide layer on the wire surface plays a significant role in achieving desired surface conditions. Without the oxide layer, tungsten wire could not be drawn due to severe breakage and plastic deformation accompanied by extreme die wear.

A certain oxide layer on the tungsten wire surface together with a smooth, nicely wetted graphite coating minimizes the coefficient of friction within the die. The best drawing conditions lead to a shiny black wire surface. Additives such as sodiummetasilicate or metaphosphate of some weight % have obviously a noticeable effect on the oxidation characteristics and stability of the dried graphite at given drawing temperatures. Lubrication, die quality and temperatures interact to give the best surface appearance and with minimized friction act against split development.

Effect Of Working And Annealing On Pore Morphology

As a sintered ingot is worked down in size, the residual pores present in the ingot become elongated[18] (Figure 2-12). Upon annealing at high temperatures, to reduce the stored energy and avoid breaks in subsequent fabrication, the elongated pores spheroidize—i.e. return to a configuration similar to the one they had in the sintered ingot. Of course, upon further working of an annealed rod, the spheroidized pores will again become elongated.

For lower degrees of deformation, RA (reduction in area) <60%, the pores are changed into an elliptical shape. At greater degrees of deformation, the pores become more lenticular, and then like needles—their length, L , becoming very long and their width, W, very small. The sequence of pore morphological changes in working AKS-doped ingots into heavy wire is depicted in Figure 2-13. Finally, above 90% RA, the needle-shaped pores completely close up. Essentially, the shape change of the pores during working and especially in wire drawing is approximately that expected if it is assumed that the volume of a pore is con-

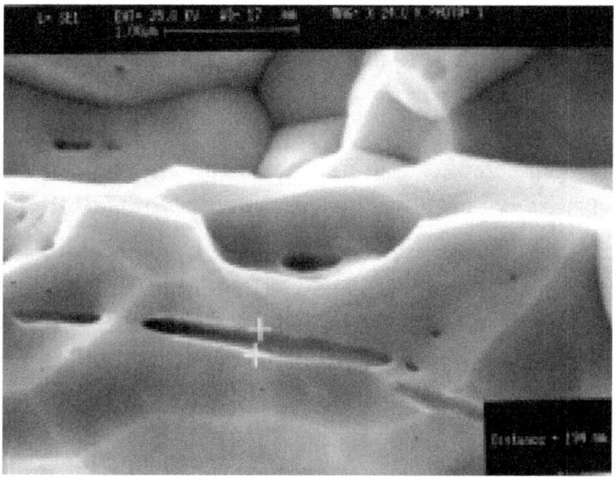

Figure 2-12. Scanning electron fractograph of 8.4mm rolled AKS-doped tungsten rod showing elongated potassium pores. (24000X.)

served during working and that the pores undergo a change in shape similar to that of the wire.

Working and annealing temperatures, notably in the later stages of heavy wire drawing, strongly influence the morphology of the deformed pores. In practice, working temperatures are reduced as the wire cross-sections become smaller and intermediate anneals are less frequently employed. When the cross-sectional reduction exceeds 70% and when the lenticular, needle-shaped pores have length to width ratios significantly >10, annealing results in two types of morphological changes depending on the annealing temperature. At low temperatures—e.g., 1000°C—expansion of the needle-shaped pores occurs. Even for fine wire sizes where no porosity could be seen before annealing, the expansion brought about by the annealing makes the pores readily visible. At elevated temperatures—e.g., 1600°C and above—each needle-shaped pore breaks up into a single row of spherical bubbles. Fractographs of drawn smaller diameter AKS-doped tungsten wires annealed at 2500°C show the formation of bubble strings, Figure 2-14. The bubble formation process is driven by capillary forces to reduce the surface area.

The number of bubbles which a needle-shaped pore produces is directly related to its L/W ratio which in turn is related to the amount of deformation by wire drawing. Considering one spherical bubble present in a rod at 5mm diameter, the number of bubbles that it would break up into at various stages of cross-sectional reductions and, if annealed at a sufficiently high temperature, can be calculated. The result is summarized in Figure 2-15.

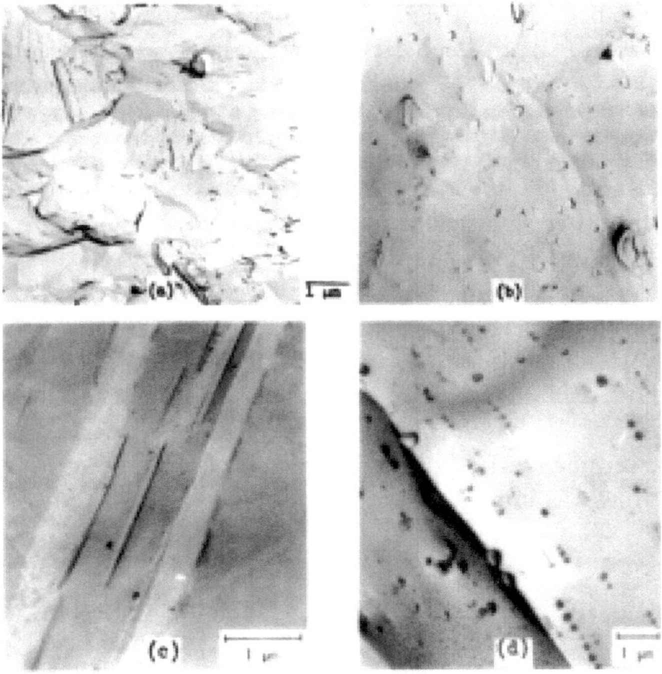

Figure 2-13. Fracture replicas of AKS doped tungsten rods and wires at various stages of processing. (a) After swaging from ingot to 11.8 mm diam. Rod, (b) same as (a) after annealing at 2500°C, (c) after wire drawing from 2.6 mm diam to 1.3 mm diam., and (d) the same as (c) after annealing at 2700°C for 20 minutes.

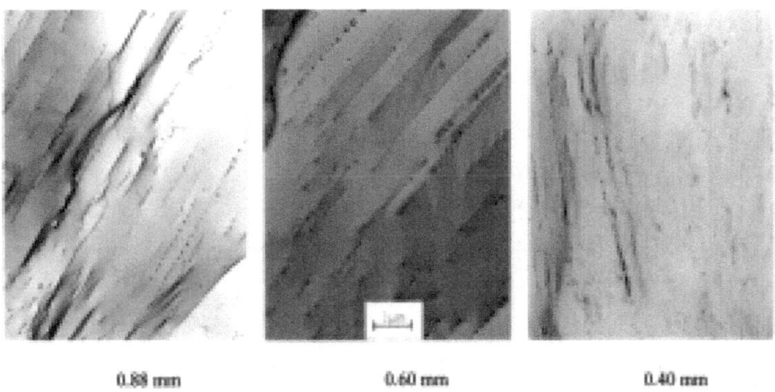

Figure 2-14. Effect of drawing and heat treatments at 2500°C for 5 minutes in vacuum on bubble size and distribution.

It can be noted that significant bubble multiplication takes place in wire of about 1.3mm diameter which has been subjected to 93% RA. Also plotted in Figure 2-15 are measured values of the length to width ratios, L/W, of the recrystallized grains for a doped tungsten rod last annealed at 4.4mm. The rapid increase in the number of bubbles, N_B, correlates well with the increase in the non-sag features of the elongated and interlocking recrystallized grains.

Annealing affects the morphological changes of the deformed pores which, in turn, affect the recrystallization process—i.e., nucleation and grain growth. The fascinating aspect of these processes is that the latter starts at temperatures of about 1000°C and above where the needle-shaped pores break up into bubble strings aligned in the drawing direction. In consequence, dislocations, subboundaries and higher angle grain boundaries become effectively pinned and especially their lateral growth is impeded (Figure 2-16).

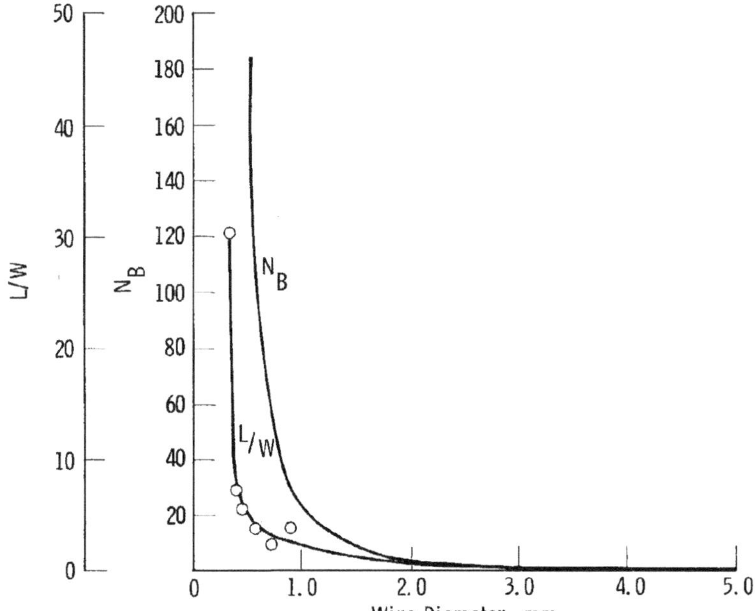

Figure 2-15. Effect of wire drawing of AKS-doped tungsten on the shape and L/W of recrystallized grains and on the number of bubbles, N_B, one pore in an annealed 4.4mm rod breaks up upon annealing at various wire sizes.

The growth of the grains is thus largely confined to growth parallel to the working direction resulting in the non-sag, interlocking grain structure (Figure 2-17).

In fact, contrary to the behavior of undoped tungsten, where the recrystallization temperature decreases with increasing deformation, as expected, the opposite is true for AKS-doped tungsten (see Figure 2-18). The bubble strings, which form upon annealing, effectively pin the grain

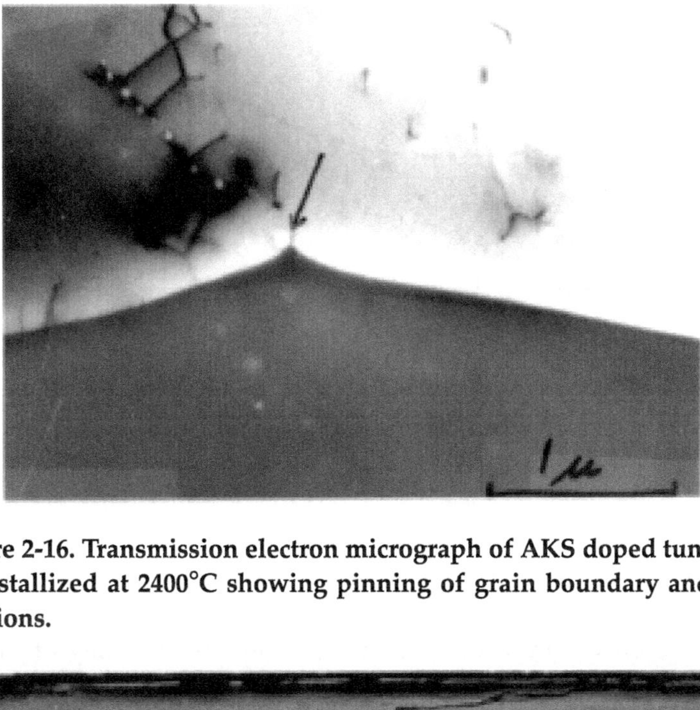

Figure 2-16. Transmission electron micrograph of AKS doped tungsten recrystallized at 2400°C showing pinning of grain boundary and dislocations.

Figure 2-17. Micrograph of 0.4mm diameter recrystallized AKS-doped tungsten wire.

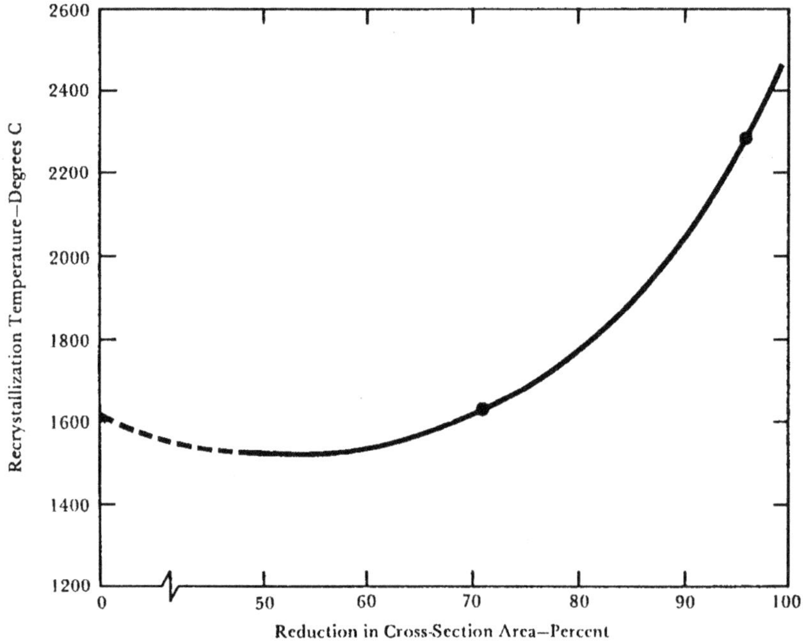

Figure 2-18. Recrystallization temperature of AKS-doped tungsten wire as a function of percent reduction in area (RA).

boundaries, reducing their mobility, so much so that greater thermal energy—i.e., higher temperature—is required to make them break away from the pins. And at this high temperature, the mobility of the few grains that do break away is so fast that they rapidly grow, consuming the smaller grains.

It is therefore important to recognize that the onset of the non-sag properties of AKS-doped tungsten requires a minimum amount of deformation following a high temperature anneal. This can be explained by the fact that at lower degrees of deformation annealing at high temperatures re-spheroidizes the pores and thereby eliminates the elongation they have already received. In essence, significantly greater deformation is required without annealing to elongate the pores into the shape of needles so that upon annealing bubble multiplication can occur and the bubbles can inhibit grain growth.

Wire Effects in Lamp Operations

Creep of Tungsten Wires

Creep resistance at high temperature is the most important attribute of the tungsten lamp filament. The early Coolidge process produced a fine ductile tungsten wire which recrystallized into a large bamboo grain structure. At the temperature of the filament (2500°C), the large bamboo grain boundaries have little resistance to deformation and slide over one another. The reduced cross-section for electrical current at these off-set locations causes local heating, which promotes greater sliding, cascading to early filament failure.

The introduction of doped tungsten wire with its overlapping grain structure seemed to prevent easy grain boundary sliding, and the coil or coiled-coil filaments had much higher creep resistance. This increased creep resistance was essential for the next important advance in filament performance, from the single coil to the coiled-coil geometry. Studies by Langmuir had shown that such geometries greatly enhance the luminous efficacy of gas filled lamps by decreasing the convective losses.[19] Maintaining this efficient geometry placed a high premium on the creep resistance of the doped wire, because the coil geometry greatly magnifies a small strain in a wire element to result in large deflections of the coil. These deflections of the coil cause sag, thereby opening up the helical turns and thus decreasing the efficiency of the coil.

This strain magnifying property of a coil was used to conduct a series of experiments to understand the factors leading to the creep strength of fine wires and to attempts to determine if the potassium bubbles played any additional role beyond that of changing the recrystallized grain shape.[20] The experiments on fine wires were important since previous investigations on creep of large specimens of pure tungsten had shown that the rate controlling process was bulk deformation by dislocation climb, rather than grain boundary sliding.[21] This was explained by the fact that the specimen diameter was much larger than the grain size. When grain boundary sliding takes place in such a situation, the stress is transferred onto the triple points and in order for creep to proceed, bulk grain deformation needs to take place. The rate controlling process for creep is then the much slower climb of dislocations.

In creep experiments on fine wires of pure tungsten, the grain size was comparable to the wire diameter and grain sliding could proceed without the need for dislocation motion which should be typical of the situation in a lamp filament. The stress dependence of this creep for the fine wire specimen (undoped tungsten, 80μm) was found to be much lower than for bulk specimens of undoped powder metallurgy tungsten (PM-W), consistent with grain boundary sliding being rate controlling (see Figure 2-19). When the wire diameter was increased to 225μm (UD-W 225μm), grain boundary sliding was again restricted and grain bulk deformation by dislocation climb became the rate controlling process, comparable to bulk PM-W (see Figure 2-19).

When experiments were then carried out on 225μm diameter AKS-doped tungsten wires,[22] it was found that they had a much greater strength than the pure wires (see Figure 2-19). Thus, even though in the large-diameter undoped tungsten wires, creep by grain boundary sliding was controlled by dislocation climb, the doped wires nevertheless were much stronger. This was surprising, since the classical explanation for the improved lamp performance of AKS-doped tungsten was that the large overlapping "non-sag" grain structure prevents easy deformation by sliding of the grain boundaries.

The remarkable strength of the doped wire appears to result from the fact that potassium bubbles are serving as effective barriers to the motion of the dislocations that control creep. AKS-doped tungsten thus takes on the characteristics of a dispersion-strengthened alloy. The fact that bubbles could impede dislocation motion has been treated theoretically[23] and was experimentally observed in studies on the mechanical properties of irradiated materials.[24] Based on the bubble density of doped tungsten, theoretical models give a reasonable agreement with the creep strength difference between pure and doped tungsten wire.

Other Implications of Potassium Bubbles in Filaments

The presence of a high density of small potassium bubbles (most smaller than 1μm) in incandescent filaments raises the question: What will happen to the bubble distribution over the life of the filaments, and will the bubbles affect the performance of the filaments? With long-time burning, the bubbles are not stationary and can be expected to move around and collide with one another. Conceptually, is is easy to see that when two bubbles collide, they will form one larger bubble, and so on. Such a process may account for the formation of the rather large holes that are often seen in burned out filaments of AKS-doped wire.

The migration of small bubbles in tungsten at operating tempera-

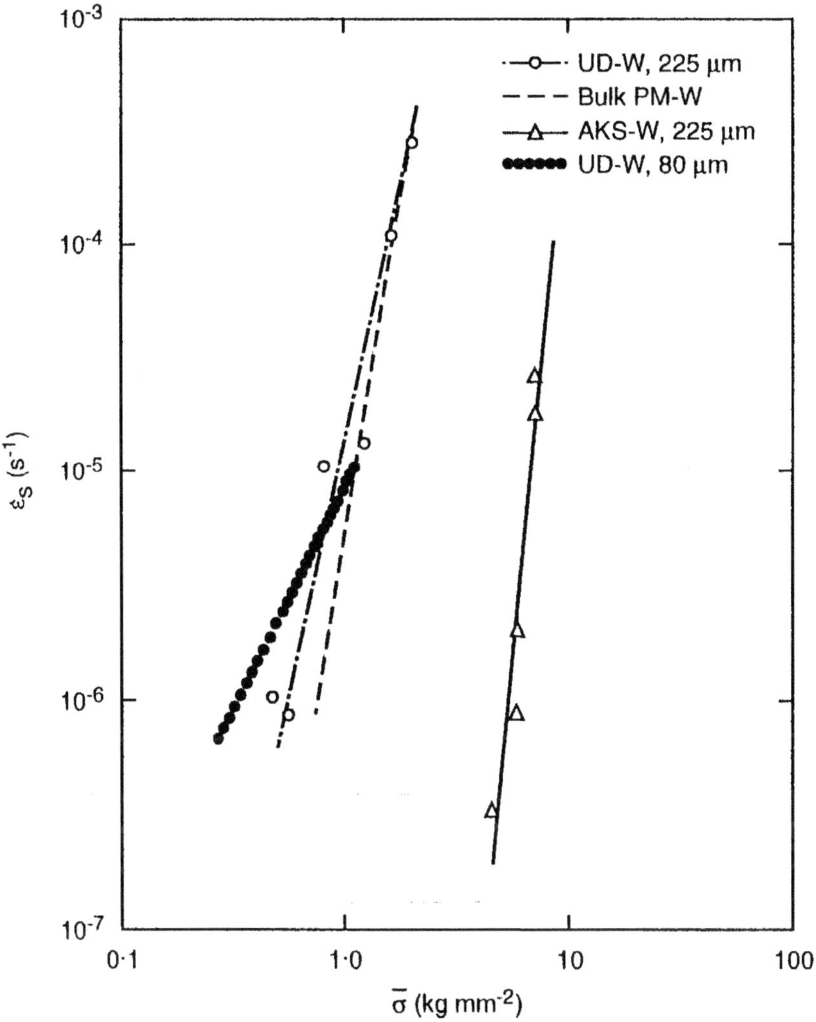

Figure 2-19. Comparative creep strength of undoped tungsten (UD-W), AKS-doped tungsten wires (AKS-W) and bulk powder metallurgical tungsten (PM-W) at 2750°C.

tures of lamps has been considered by Sekerka[25] with respect to their behavior in a uniform temperature environment and under the condition of small variations in filament temperature. He found that for the bubble sizes normally encountered in AKS-doped wire, and in a uniform temperature environment, the bubbles would migrate around in a random manner by the mechanism of surface diffusion. In a temperature gradient, on the other hand, their motion will be biased with the bubbles migrating toward the hotter regions. This effect was confirmed by Dawson[26] who demonstrated that in a thermal gradient, bubbles coalesce much faster than in a uniform temperature, with the bubbles growing in the hotter section. The coalescence of bubbles up the temperature gradient to the hotter sections must raise the resistance of this section of the wire so that it becomes hotter, causing the development of a hot spot causing eventual filament failure.

From this result, Sekerka has developed a linearized theory of filament failure taking into account the instability of bubble density and temperature in the presence of small perturbations:

$$<r>^n = At\frac{mkTa_o^4}{Y}e^{Q/RT}$$

$<r>$ = root mean square bubble radius
t = time,
Y = surface tension = 2500 mgs/cm^2
Q = activation energy
R = gas constant = 1.98 cal/°K/gmatom
T = temperature
m = number of gas atoms per unit volume 2×10^{19} atom/cm^3
k = Boltzmann's Constant 1.38×10^{-16} mg/°K
a_o = lattice parameter = 3.165×10^{-8} cm

The validity of this equation was confirmed by examining the bubble coalescence in 8mm wire by burning wire at 2800°C and measuring the change in bubble size distribution with time.

The experimental results agree with a value of $n>1/5$ and a value of Q of approximately 70Kcal. In the course of these experiments and those of Dawson it was noted that bubble coalescence on the grain boundaries occurs much faster than in the grain interior (see Figure 2-20).

The fact that good lamp wire has an abnormally large grain size is fortunate and no doubt reduces the kinetics of bubble coalescence.

Of course, one is tempted to substitute other alkali metals for potassium, such as lithium or sodium, since these elements have similar properties. Preliminary experiments with dopants of Li and Na were conducted.[27] The results were not encouraging. In every material tested, the bubble coalescence was more pronounced than in AKS-doped tung-

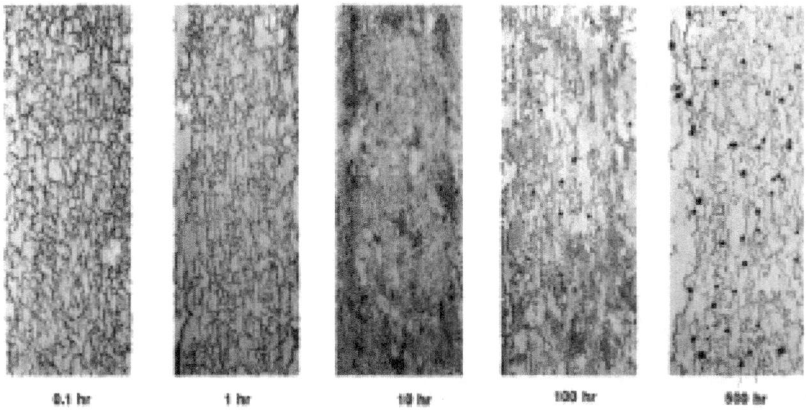

Figure 2-20. Effect of burning time at 2800°C on the bubble distribution in 0.06mm AKS-doped tungsten wire (200X).

sten, with one exception. A wire doped with the standard dopants, but treated to reduce the aluminum to a very low level (<10 ppm), exhibited coalescence that was lower than in AKS-doped tungsten. This result suggests that Al is an important impurity affecting the rate of coalescence and that steps taken to eliminate Al from the sintered ingot should be beneficial to filament life.

Conclusions

As a result of the work presented in this review, bubbles are seen to play a dominant role in the performance of incandescent lamp filaments. The bubbles appear first during sintering. This and every other subsequent processing step given to the material alters the bubble distribution in such a way that a non-sag filament results at fine wire sizes.

Naturally, it challenges the imagination that Pacz's invention[5] of a creep-resistant lamp filament appears to have been a combination of rather startling events, totally unknown to the inventor. Starting with a strange combination of dopants, he could never have suspected that these dopants would chemically react during subsequent reduction and sintering to trap the highly volatile impurity potassium within the tungsten lattice. Equally unpredictable must have been the fact that this volatile impurity would create bubbles in the wire which, if processed correctly by avoiding intermediate anneals during wire drawing, would multiply and result in a large, near single crystal grain structure, with grain boundaries at a high angle to the wire axis.

The final serendipitous event in this amazing string of unique events was that just getting rid of the bamboo grain structure of pure, undoped tungsten did not explain the very high creep strength of the AKS-doped wires. The high strength resulted instead from the fact that the then unknown bubbles are very effectively pinning dislocations, the defect controlling creep deformation mechanism that was not discovered until nearly 30 years later.

Future Outlook

Three major factors will exert their future influence upon the production of tungsten wire and rods for lamp applications: 1) the implementation of the concepts of "Total Quality Management" (TQM), 2) the ongoing quest for energy conservation in lighting and 3) the continuing growth of the global market for light sources at a rate of 3-4%.

Factor 1, TQM, refers, in part, to "statistical process control" combined with quality cost analysis. The latter, of course, is related to process costs which are the driving force for cost reduction measures. The quality of all lamps depends to a very large extent on the quality of tungsten—i.e. the uniformity and consistency of the doped tungsten wire and rods. Significant improvements in doped tungsten wire quality must be anticipated as it may be sprung onto the industry by a hitherto unknown source.

Factor 2, energy conservation, will continue to create the pressure for the development of more energy-efficient lamps. In turn, the tungsten wire manufactures will be called upon to supply tungsten with modified metallurgical, surface and compositional properties. They can expect to face enormous challenges to supply the lamp manufacturing plants with new and different tungsten wires and filaments, mostly on short time notice. Among the lamps which will receive continued attention, are standard incandescent lamps, because the life of incandescent lamps is not yet fully optimized and improving failure distributions demand tungsten wire of greater uniformity. Research on substitute incandescent materials with the promise of higher luminous efficacies and the potential for energy conservation, such as tantalum carbide or refractory oxides, may be reinitiated, although there is presently little expectation for a commercial breakthrough.

Factor 3, global market growth, is bound to impact on manufacturing capacity. Although electrodeless, RF-energized lamps are under development, they are not likely to affect the demand for tungsten wires and rods in any major way for many years to come. The manufacturers will have to deal with ISO 9000 and with lamp manufacturers in other countries. Anticipated expansion of manufacturing capacity will provide the impetus for the design and installation of technologically superior manufacturing equipment.

References

1 P.W. Keating. *Lamps for a Brighter America.* New York: Mc Graw-Hll Book Co., New York, 1954.

2 *The American Heritage Dictionary,* Third Edition. New York: Houghton Mifflin Company.

3 H. Mennike. *Die Metallurgie des Wolframs.* Berlin: M. Krayn Verlag, Berlin, 1911.

4 W.D. Coolidge, Trans AIEE, [1910], 961 and US Patent No. 1, 082, 933, [1912].

5 A. Pacz, US Patent No. 1, 410, 499 , [1922].

6 L. Bartha, A. Kiss, J. Neugebauer, T. Nemeth, HTHP14 [1982] 1.10.

7 H.J. Lunk, "Investigation of Powder Metallurgical Production of Doped Tungsten with Unorthodox Doping Specification," *Int. J. of Refractory Metals and Hardmetals,* 14 [1996] 271-277.

8 H.J. Lunk, Neue Hutte 27, Nr. 9 [1982] 329-334.

9 H.J. Lunk, P.Schade, Neue Hutte 27, Nr. 10[1982] 382-387.

10 A. Gahn, "Bubbles" Dissertation, Ludwig Maximilian Univ. Munich, [1986].

11 T.R. Wilken, W.R. Morcan et al., Met. Trans, B 7 [1976] 589-597.

12 H. Zeilinger, Diplom Thesis, Ludwig Maximilian Univ. Munich, [1985].

13 D.M. Moon, R.Stickler, A.L. Wolfe: Proc. 6th. Plansee Seminar, Ed. Benesoosky, F. Springer, Vienna-New York, [1969] 67.

14 A. Gahn et al: own investigation.

15 H.G. Sell, D.F. Stein, R. Stickler et al., *J. of the Inst. of Metals,* Vol 100 [1972] 275.

16 M. Lee, D.G. Flam, "Lubrication Mechanisms in Tungsten and

Molybdenum Wire Drawing," *Lubrication Engg.* 47 [1991] 2, 127 ff.
17 W. Martens, P. Brulez, "Aspects of the Constitution of Tungsten Wire Surfaces and itís Relation with Friction in the Wire Drawing Process," 12. Proc. Plansee Seminar,Vol 1, Ed. Bildstein v. Ortner, paper RM 50 [1989] 523-537.
18 D.M. Moon, R.C. Koo, *Metal Trans.* Vol 2 [1971] 2115.
19 Langmuir, I, US Patent No. 1, 180, 159 [1916].
20 D.M. Moon, R. Stickler, *Phil.Mag.* Vol 24 [1971] 1087.
21 H.G. Sell. *Refractory Metal Alloys.* Ed.: I.Machlin et al. New York: Plenum Press, 1968, p. 395.
22 D.M. Moon, R.Stickler, *High Temperatures-High Pressures,* Vol 3 [1971] 503-516.
23 P. Coulomb, *Acta Met.,* 7 [1959] 556.
24 J. Moteff, R.C.Rau, F.D. Kingsbury, *Symposium on Radiation Damage in Reactor Materials,* Vienna [1969].
25 R.F. Sekerka, *Westinghouse Research Report* [1967].
26 C.W. Dawson, *Met. Trans.* 3 [1972] 3103.
27 D.M. Moon, Experimental Results, private communications, [1996].

Chapter 3

The T3 Quartz Infrared Lamps

by Raymond Kane and Steve Sirek

Heat Sources

The development of better radiant infrared heat sources was driven by the desire to generate higher heating levels for industrial, commercial and even military applications. The early heat sources were hollow metal tubes containing a coil of nichrome, or similar high-resistance wire, that produced a peak radiation at a color temperature of 925K. When that same coil was placed in a quartz tube, the re-radiated heat from the quartz raised the peak radiation temperature to 1250K. In contrast, infrared incandescent lamps peak at 2500K.

From the Stefan-Boltzman Law, we know that for an infrared source, the radiant energy output is directly proportional to the fourth power of its absolute temperature. From this, we can see that the incandescent heat lamp for a given wattage is capable of delivering much more usable heat than the nichrome coil sheath. For, as the radiant peak temperature increases, the greater proportion of the infrared energy produced is under 5 microns[1] (see Figure 3-1). Most of the radiant energy over 5 microns is absorbed by the envelope, so it is not available for supplying heat to any target.

Raising the peak temperature to 2500K shifts the energy distribution curve closer to the visible portion of the spectrum. The peak becomes more pronounced and the radiant energy that is emitted drops off rapidly to about 0.8 microns, which is still above the visible limit of 0.75 microns. Most of its infrared energy is in the efficient utilization region—under 5 microns. The amount of energy that is seen as light in the infrared lamp is about 15%, and most of the remaining 85% is transmitted to the target as heat.

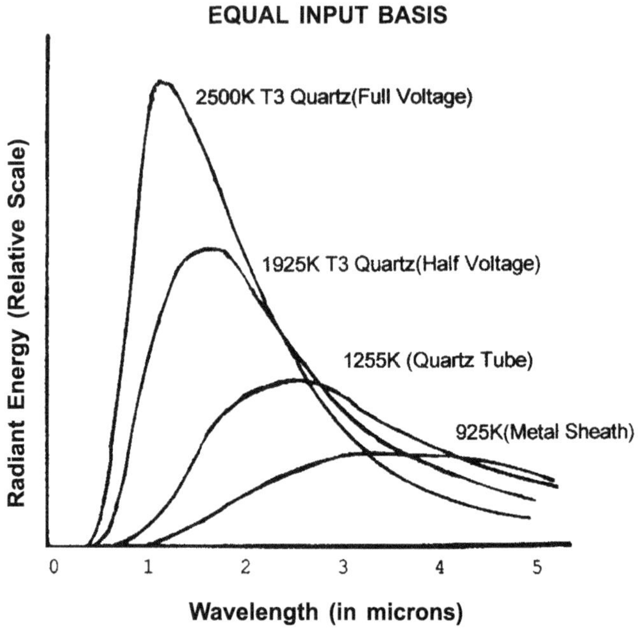

Figure 3-1. Family of spectral energy curves showing infrared radiation from various sources. Courtesy: Fostoria Industries, Inc.

There are several features that make the T3 quartz infrared lamp a unique heat source. Its small size reduces the necessary size of its mounting fixture. Its linear design provides a uniform distribution of heat, especially in conveyor ovens. Its quartz envelope can handle wide thermal changes, and when hot is resistant to breakage from water and other solvents.

The Development and Construction of T3 Quartz Infrared Lamps

Conventional Incandescent Lamp Bulbs

The original infrared lamps designed for use in industry and the home were made with conventional incandescent lamp bulbs, such as the reflector types (R40) and the globular types (G30). Most of the lamps made for home use were 250W R40 lamps that used a red bulb to reduce glare and generate the feeling of "warmth."

Lamps for industrial use were made in the 125/250/375/500W sizes and almost always in clear, hard glass bulbs.

The problem with the lamps for industrial use was that their circular beam patterns did not provide the uniformity of heat that was generally required in linear conveyors and ovens. In addition, the reflectors that were used were very bulky and could not focus the heat uniformly to all parts of the product that were being baked or dried. Furthermore, and even after hard borosilicate glass was in use, the power density of the lamps was still relatively low, with maximum levels of only 1000W/sq.ft.

Tubular T3 Quartz Lamps

When the tubular T3 quartz lamps were introduced in the late 1950s, with their compactness, focusing ability and high thermal strength, it became possible to achieve power levels of up to 5000W/sq.ft. Foote and Hodge of General Electric are credited with the original design.[2]

The T3 lamps are made with either a tube of relatively inexpensive translucent quartz which is obtained from high-quality silica sand, or a more costly tube of clear fused quartz that comes from crystals of high purity silicon dioxide. The translucent bulb spreads out the heat from its diffuse surface and reduces the glare from the visible light. The clear bulb radiates a sharp, bright source that can be precisely directed on the item being heated. Translucent quartz, with its greater heat absorption, cannot be operated at as high a temperature as the clear bulb, because it will soften.

End Terminal Designs

The electrical connections on the lamp ends are generally nickel pigtail wires covered with a nickel sleeve, as shown in Figure 3-2a, or else a silver-plated nickel recessed contact button, set inside of a ceramic base as shown in Figure 3-2b. This button terminal design also provides for a more precise positioning of the lamp when used in optically designed reflectors. Other base configurations have been designed for special applications.

Other Design Features

Usually filled with about 600mm of argon gas, the lamp operates

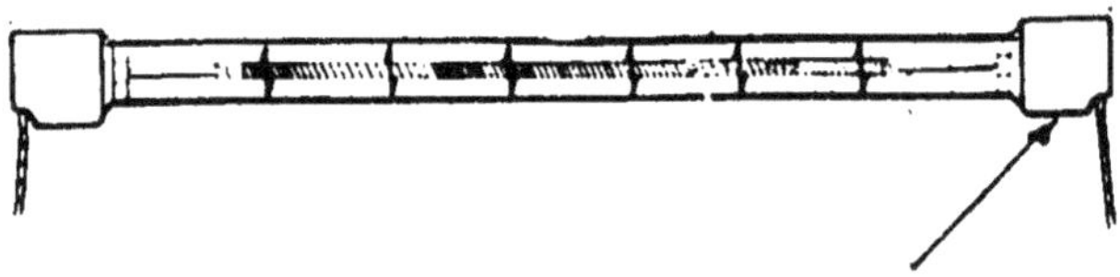

Figure 3-2a. T3 quartz lamp end terminals constructed of a nickel sleeve with stranded wire leads. Courtesy of Philips Lighting Company.

Figure 3-2b. T3 quartz lamp end terminals constructed of a ceramic base covering a recessed contact button. Courtesy of Philips Lighting Company.

close to atmospheric pressure. With its long, linear filament in close proximity to the bulb wall, it would be quite possible to blacken the wall from tungsten evaporation. Fortunately, the infrared lamp normally requires a low operating temperature of the tungsten, which reduces evaporation and extends its life to well beyond 5,000 hours. Furthermore, the disks or wires that support the filament at regular intervals along its length are made of tantalum, which is a good getter for hydrogen and oxygen, two components that encourage tungsten migration by the so-called "water cycle."

Most T3 infrared lamps are designed to operate at 100W per inch, so the total wattage can be determined by measuring the overall length of the filament. However, some high temperature infrared lamps have been designed to operate at 200 W per inch.

Burning Position

As might be expected with a long filament that is supported by thin disks or spirals of wire, the infrared lamps are typically designed just for horizontal operation. Overall lamp lengths can range from 8 inches, as in the 300W size, up to 42 inches, as in the 3800W size. In

1993, Philips Lighting made some lamps that were 6 ft. long!

Burning the lamp in the vertical position will cause the heavy filament to collapse from its own weight, making it sag and short-out its windings. It will usually then fall against the bulb wall, softening and bulging the quartz to the point of lamp failure.

Lamps have been made for vertical operation by melting small indents at opposing points in the quartz bulb, just under each disk or spiral support (see Figure 3-3). This capture of the support by the quartz limits the length of filament that can collapse, which minimizes its "catenary" sagging.

Seal Temperature Limits

When used in very high power density applications, it is absolutely necessary to keep the seal temperature below 350°C to prevent the oxidation of the moly seal ribbon. One way this can be done is by isolating the seal ends in confining partitions built within the fixture, then blowing cooling air over them. Sometimes, the ends of the lamp are bent at a right angle, so that they can protrude outside of the oven or reflector to keep the end seals cool.

Another method is to use platinum-coated (or even pure platinum) outer leads that are then welded directly to the ribbon. Another is sealing the entire exposed press area of the quartz end with a glass frit or chloride and borate salts that will not melt at the elevated operating temperatures. Figure 3-4 shows a T3 quartz lamp with a platinum protected end seal, further protected with lead borate.[3]

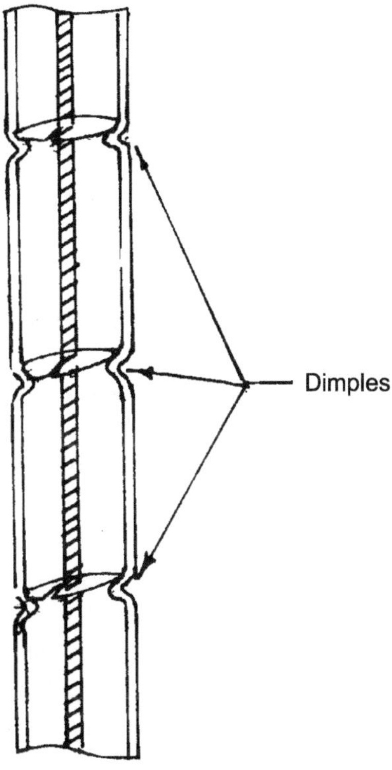

Figure 3-3. Recessed "dimples" melted into the quartz tubing to support the tantalum disks to permit vertical burning operation.

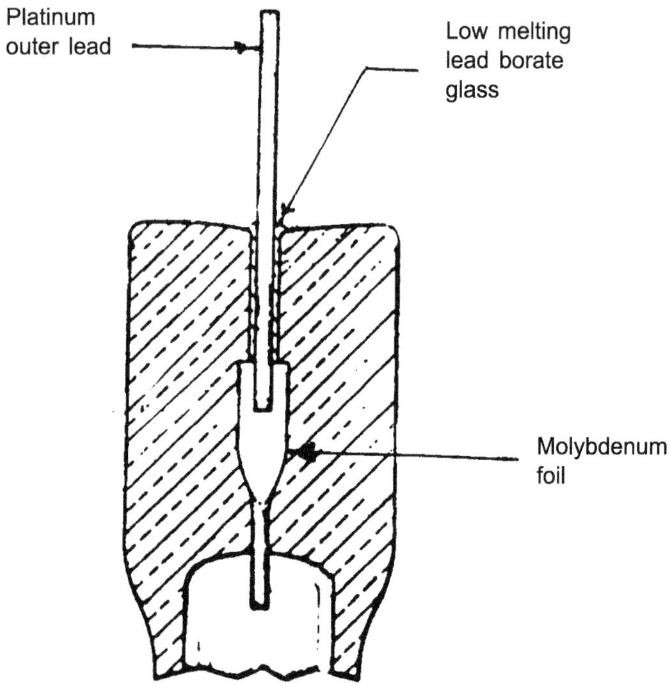

Figure 3-4. An end seal containing a platinum terminal wire and lead borate filler to protect the moly ribbon. Source: Holcomb, R.H. and Thomasson, G.I., Quartz To Metal Seal, U.S. Patent 3,211,826, 1961.

The Tipless Lamp

One of the problems in directly focusing the infrared energy to the exact work location was the presence of the exhaust tip in the quartz bulb, at the point where the lamp is evacuated and filled with argon. The radiant energy passing through the tip would distort the precise thermal linear pattern required on the workpiece.

In the early 1970s, Sylvania, followed by others, developed a line of tipless infrared lamps. Previously, the lamps were made by flowing a flushing gas of nitrogen or argon through the side exhaust tube while the first end is sealed. Then the bulb was turned over and the second end was sealed. The sealed body was transferred to an exhaust machine where the residual gases were removed and the argon fill gas was added. The exhaust tube was then tipped off and the lamp was completed.

Sylvania used the pure argon fill gas for flushing, then made the first pressed seal, and about one second later, made the second seal with a second set of press jaws that were located at the opposite end of the press machine. The other lamp companies developed similar methods to produce tipless, distortion-free lamps.

Vycor Bulbs

Another development involved bulbs made of vycor, which is 96 percent silica. In the pre-fired state, Vycor is very porous. At that point it is possible to impregnate it with various solutions to achieve different colors. For infrared lamps, it was logical to have tubes that would be deep red in color.

In 1963, Corning developed such tubes, by impregnating them with an iron-nickel compound.[4] Subsequent baking caused the vycor to shrink and trapped the red compound uniformly inside. The ends, however, had to be kept clear, otherwise the sealing fires would be reflected away from the ends and the vycor would not soften enough to be pressed. This meant that each bulb had to be cut to the correct lamp length before impregnation and left with about one-half inch of the ends clear. Allowance had to be made for the subsequent shrinkage.

The Halogen Infrared Lamp

With the advent of the space program, no amount of infrared radiation produced by these lamps seemed to be enough to provide the simulated high temperature, re-entry conditions needed to test the outer surfaces of space vehicles.

Lamp wattages were pushed to their limits by greatly raising the lamps' operating voltages, with a consequential drastic reduction in lamp life. In spite of extensive measures taken to improve the getters, remove more of the gaseous contaminants from the parts, and use highly purified tungsten filaments, the bulb walls, nevertheless, quickly blackened and failures occurred. During life, the irregular blackening was also distorting the optical pattern of the infrared radiation.

That was the impetus for the development of the halogen infrared lamp. The halogen, as iodine, was originally introduced as a getter, because as a scavenger, it readily combined with oxygen, hydrogen and carbon. Soon it was realized that a "halogen transport cycle" was keeping the tungsten from reaching and depositing on the bulb wall. Extensive testing followed to establish the operating temperatures required to optimize this sought-after transport cycle.

Furthermore, it was found that the higher filament color temperature of the halogen lamp at 3000K improved the radiant efficiency over that of the standard T3 quartz heat lamp which operates at 2500K.

Applications for T3 Quartz Infrared Lamps

As a source of providing infrared radiant energy, T3 quartz tungsten filament lamps have become popular for many reasons. They are relatively small and inexpensive, provide heat that is clean and uncontaminated, and can be made in a wide variety of wattages, lengths and loadings. Their energy can be directed with reflectors that are highly polished aluminum or even gold-plated, to make them perform very efficiently.

Linear quartz infrared lamps as compared with their globular predecessors, permit the use of smaller ovens, provide a more uniform heat distribution and allow great flexibility in modifying conveyor temperatures.

Applications In The Space Program

Probably the biggest boost given to the popularity of the linear quartz lamps was the United States space program, where it was found that these lamps could be packed tightly together and over-loaded, at least for short periods of time, to simulate the intense heat generated when the space vehicles re-enter Earth's atmosphere.[5] Lamps rated at 2.5kW can be loaded, for brief periods, to 15kW, then clustered to deliver power densities as high as 28kW/sq.ft.

The materials used for the frame structure, the skin and the ablative coating of these vehicles must not only perform well and reliably at these temperatures, but must do so in the near-vacuum conditions of space!

Other Applications

These versatile lamps are used in the following types of heating applications, with new ones being found every day:

Food Processing - The low coefficient of expansion of the quartz provides a safety factor. Glass lamps can break from grease and water splatter.

Chemical Processing

Where explosions might occur if open flames had to be used, the intensity of the heat can be altered by simply adjusting the voltage to the lamps.

Production Paint Lines

The ability to control the heat precisely allows for even drying of painted surfaces, even those that are irregular in shape and cumbersome to handle, such as automobile body parts and even entire railroad cars.[6]

Thawing

Infrared heat is also used to thaw the loads of frozen coal and ore shipped in railroad cars, prior to dumping their contents.

Conveyor Ovens

The linear lamp design provides the uniform heating needed to properly bond adhesives to carpeting, to dry inks and fabrics, to fuse lettering on glass and for pre-heating plastics and wood prior to shaping and forming.

Outdoors And Other Cold Areas

The radiant heat directly warms the people instead of the surrounding air. Installations include bus stops, airports, train depots and entrances to stores, where they serve another function of melting the snow near the doors.

Incubation In Modern Hatcheries

The uniform heat is very beneficial to the well-being of the poultry.

Copying Machines

The combination of even heat and uniform light supplied by the linear lamps produce clear, high-resolution documents. Half-tones and pictorial contrasts can be obtained by adjusting the lamp voltage.

In spite of their higher cost compared with other heat-producing sources, the versatile T3 quartz infrared lamps for many years now have dominated the fields of heating, drying, baking, curing and testing a vast range of materials, products and components.

References

1 Barber, .I.J., *Rules of Radiation*, Fostoria Pressed Steel Corp., Brochure 58-120, 1958.

2 Foote ,A.G. and Hodge, W.F., "Infrared Ray Generating Device," U.S. Patent 2,864,025, 1958.

3 Holcomb, R.H. and Thomasson, G.I., "Quartz to Metal Seal," U.S. Patent 3,211,826, 1961.

4 Elmer, T.H. and Nordberg, M.E., "Lamp Having a Colored Bulb," U.S. Patent 3,258,631, 1966.

5 Hodge, W.F., "Developments in Quartz Infrared Lamps", Illuminating Engineering, 56-1: p. 333, 1961.

6 General Electric Company, *Product Heating with Infrared Lamps*, Brochure TP-116-R, 1964.

Chapter 4

The Tungsten Halogen Lamp

By Steve Sirek and Raymond Kane

A Brief History

The concept of using a halogen in a lamp was first patented in 1882.[1] It stated that the addition of chlorine vapor to a carbon lamp would improve its efficiency by reducing the wall blackening. In 1894, Professor William A. Anthony reported similar success using bromine.[1] While this knowledge was a laboratory curiosity, there was no practical use of this information for many years. Then, in 1933, J.A.M. Van Liempt patented the process for using the halogen, iodine, to capture the tungsten evaporated from a glowing filament and by vapor transport, return it to the filament.[3]

Unfortunately, the patent did not lead immediately to a practical halogen lamp. There were problems with filament ends and supports being attacked by the halogen. Excessive amounts of tungsten that deposited in localized regions of the filament resulted in short lamp life. Nonetheless, there was a continuing effort to refine the tungsten-iodine reaction, as thoroughly documented by Moore and Jolly.[4] Iodine was also used for arc suppression, even to the point of being able to eliminate the conventional lamp arc suppressant, nitrogen.

The Halogen Cycle

In 1959, Fridrich and Wiley of General Electric were granted a patent for a design that lead to the first commercial tungsten-halogen lamp.[5] It contained iodine.

The iodine was added to a tubular quartz infrared lamp to control the blackening in these small, but highly loaded, lamps that had been introduced only a few years before. They were being used to produce intense heat to simulate the re-entry of space vehicles returning to earth from orbit.

Fridrich and Wiley are credited with introducing a small but precise amount of iodine vapor into the lamp which greatly enhanced its performance. Others, including Zubler and Mosby at General Electric,[6] Schilling of Osram[7] and Van Tijen of Philips[8] worked on the problem diligently, and contributed much to the new technology.

Other companies, among them Westinghouse, Sylvania and GEC of England, were similarly investigating this intriguing halogen cycle. By 1970, halogen lamps had expanded to a wide variety of designs from the simple linear lamp first introduced in 1959.

There were many types and sizes of linear halogen lamps that contained iodine crystals. Other lamps were introduced that used bromine gas. It was found that dosing bromine in with the fill gas made production much easier than trying to control the method of getting an unpredictable amount of iodine vapor into the lamp. Many iodine lamps were soon redesigned to use either bromine or a gaseous iodine compound such as methyl iodide.

Controlling The Fill Pressure

The first halogen lamp designs were much like their incandescent counterparts, filled to a gas pressure of about 600 torr when the lamp was cold. The halogen lamp's strong quartz envelope, however, made it possible for it to operate well above atmospheric pressure.

With quartz envelopes, liquid nitrogen can be used to condense the fill gas during the filling operation. When the lamp returns to room temperature, the gas can be at a pressure of least 2 atmospheres. The higher fill pressure reduces tungsten evaporation significantly, as does replacing nitrogen with heavier gases such as argon, krypton or xenon.

Eventually, lamps were being filled to as much as 15 times atmospheric pressure at room temperature. When operating, the heated lamp's pressure can reach up to five times the value of a cold lamp, although not quite as much for lamps whose volume is greater than 0.4 cc.

The Introduction of Hard Glass Bulbs

In the 1980s, halogen lamp technology expanded into hard glass envelope designs. It was discovered that borosilicate glasses could withstand the thermal shock of liquid nitrogen and could be cold-filled to as much as 5 atmospheres.

The first hard glass halogen lamps were made with PAR outer bulbs. Their thick borosilicate envelope protected the consumer if the halogen lamp inside were to fail violently.

Reflective Coated Bulbs

The late '80s and early '90s saw the introduction of halogen lamps made with an infrared reflective coating that increased the filament temperature and improved the lumen efficacy of the lamps.

When halogen lamps were introduced in 1959, they had a typical incandescent efficacy of 15 lumens per watt (LPW), and now IR-coated lamps deliver 28 LPW while at the same time increasing the life from 750 hours to 3,000 hours.

The Need for Higher Efficacy and Smaller Sources

The incandescent lamp has several advantages: It can operate without the need of auxiliary equipment such as ballasts and starting ignitors; it can operate in a wide range of temperatures; it is a comparatively small light source; and it is low in cost. Its disadvantages are its poor lumen maintenance, its relatively short life and its low efficacy.

With only 15 percent of its radiant energy in the visible spectrum with the other 85 percent in the infrared region, this lamp is a better heat source than a light source.

The late 1950s and the early 1960s were periods of great technological achievements for halogen lamps. The space program, with its great need to develop compact and light-weight equipment, led to a similar desire with the general public to want smaller and lighter appliances and other products, including lamps.

General lighting lamps were being designed to be smaller and more efficient. Many of the early halogen lamps were higher efficacy modifications of linear iodine infrared lamps. Figure 4-1 shows the difference in size between a 1500W halogen lamp and its predecessor, the standard 1500W lamp. Further compactness of the linear lamps led to the development of short single-ended lamps for general illumination.

Similar compact, high brightness lamps soon found their way into

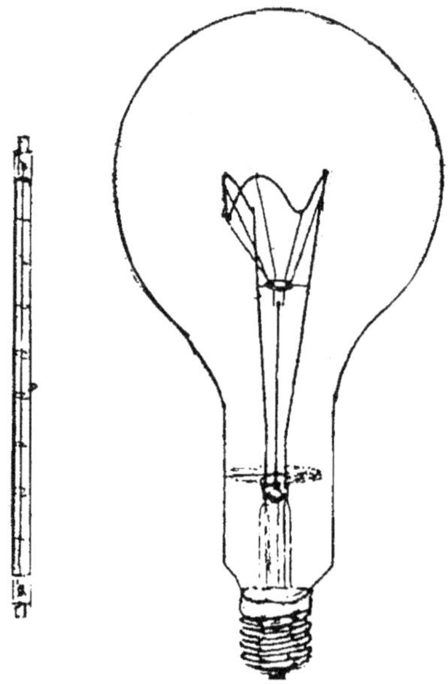

Figure 4-1. Size comparison between a 1500W halogen lamp and a 1500W incandescent lamp (from Strange & Stewart, 1963). Courtesy of the Illuminating Engineering Society.

other special lighting applications such as slide projectors, magnification systems, automotive and aviation lighting.

The Development of Tungsten Halogen Sources

The basic action of the halogen lamp is its ability to produce a tungsten regenerative cycle which takes place when the evaporated tungsten leaves, then returns, to the filament without ever condensing on the relatively cool bulb wall.

Between the filament and the bulb wall, there are three significant temperature regions of interest (see Figure 4-2). The first region, "A," is where the tungsten atoms that have evaporated from the hot filament are moving toward the bulb wall, where they will combine with iodine atoms to form tungsten iodide. The bulb wall temperature must be between 250°C and 600°C for the reaction to work, and it is where most of the tungsten atoms that evaporated from the hot filament will combine with the iodine atoms to form stable tungsten halides. In the middle region, "B," between 600°C and 1600°C, most reactions are reversible.

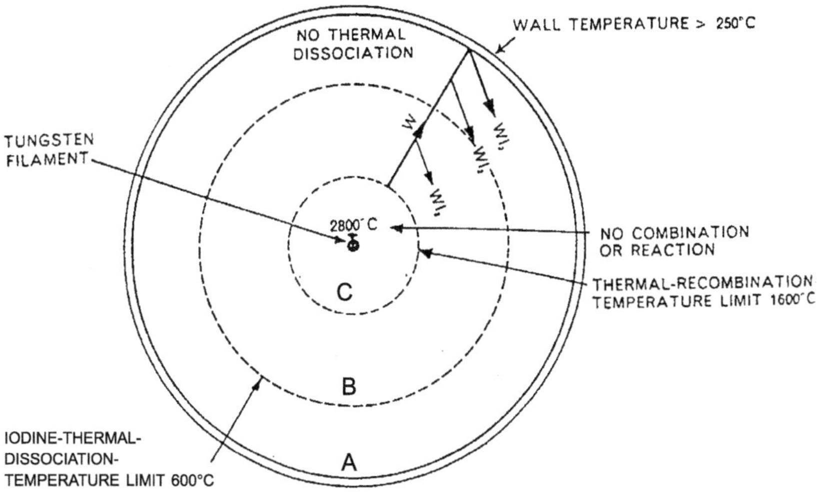

Figure 4-2. Between the filament and the bulb wall of an iodine lamp, there are three significant temperature zones of interest.

There, the tungsten iodines are able to recombine or decompose. In the core region, "C," with temperatures ranging from about 1600°C to the filament temperature of about 2800°C, the iodides are forced to disassociate into tungsten and iodine atoms.

Some of the tungsten atoms will migrate to form tungsten iodide again, but most of the others will redeposit on the filament. The cycle works similarly with bromine, chlorine and at least theoretically, with fluorine, but at differing recombination and disassociation temperatures.

Filament Failure

The incandescent lamp filament generally fails at its hottest point because that is where more of the tungsten has evaporated. The smaller cross-section for current flow, makes it even hotter, until it finally melts and breaks at that point.

The time of greatest current flow is when the lamp is first turned on due to the high in-rush current into the cold, lower resistance of the wire. The halogen lamp, like the incandescent lamp, is most likely to fail at this time. The tungsten that is deposited on the filament is not selectively attracted to the hot spot (site of failure), so lamp life is not extended by the tungsten regenerative cycle.

Halogen lamps achieve their longer life from the higher fill pressures that are used, which reduces the rate at which the tungsten is evaporated from the filament. There are three ways that the reduced rate of evaporation can be utilized:

1. A lamp designed to the same efficacy will have a longer life because of the reduced evaporation rate.
2. A lamp designed to the same life can have the filament modified to produce more light output that the slower evaporation rate of the tungsten will provide.
3. A compromise can be made between the two extremes above, to produce a lamp with both longer life and higher output.

The halogen lamps are perceived to be brighter and last longer, primarily because the comparison is made with the incandescent lamps they replace.

Halogen Dosing

The earliest halogen lamps were dosed with iodine. Crystals of pure iodine were dropped directly into the lamp and then gently heated so as to distribute the iodine along the bulb wall. The iodine must be of the highest purity because the contaminants it contains or absorbs will be released inside of the lamp as soon as the crystals are vaporized. Later on, the iodine was introduced into the lamp as a vapor, along with the fill gas which could be nitrogen, argon, krypton or some combination of the three.

Placing iodine into the lamp by vaporizing the crystals was not reliable. Excessive heat allowed too much iodine into the lamp and reduced the light output because the vapor, which is pink in color, absorbed some of the light being produced by the filament. Insufficient heat produced too little iodine vapor and the lamp blackened. Fortunately, the iodine cycle works over a broad range of iodine content. Other halogens, on the other hand, require much more critical control. Today, programmable controllers and temperature sensors provide precision dosing.

Vertical Burning

Linear halogen lamps work well when they operate in a horizontal position, with a deviation of no more that 15 degrees. Vertical burning causes the heavy iodine vapor to rest on the bottom, while the lighter inert fill gas rises to the top. Without iodine present, the top quickly blackens. English and Zubler have shown that vertical operation of the lamp can be improved by increasing the fill pressure.[9] They recommend pressures of 1500 torr for xenon, 2500 torr for krypton and 3500 torr for argon.

The Advantages of Bromine

Iodine lamps were found to require a small amount of oxygen to improve the action of the halogen cycle, although the mechanism is not fully understood.[10] Without adequate oxygen, the bulb wall blackens rapidly, a condition that also occurs if water or water vapor are present.

It would seem that lamps containing bromine would also blacken without oxygen, but that is not the case.

Bromine is usually added as a compound such as menthyl bromide (CH_2Br_2), menthylene bromide (CH_3Br), or hydrogen bromide (HBr). Dosing with menthyl bromide or methylene bromide is less critical than with hydrogen bromide, and it is believed that the carbon ties up some of the free bromide.

Dosing is a lot simpler with bromine than with iodine because the bromine compounds that are used are in a gaseous state at room temperature. They can be pre-mixed with the fill gas before dosed into the lamp. The fill gases containing the Charlie Browns (the name given to bromine compounds of carbon and hydrogen) will tend to settle in cold weather. The gas tanks, which must be at room temperature, have to be rolled periodically to remix the gases inside. Tank pressure must be kept relatively low, although the hydrogen bromide mixture can tolerate a higher pressure than the others.

Because of the highly corrosive nature of bromine, it is used with extreme caution in lamp exhaust machines, especially when dosed as hydrogen bromide. With proper care and frequent machine maintenance, however, it can be used satisfactorily. Hydrogen bromide is often the preferred dose because the others contain carbon, which embrittles the tungsten coils. However, each compound has its own advantages and disadvantages.

Higher Fill Pressures

In the early 1970s, the lamp were made with progressively higher fill pressures, some as high as 11 atmospheres. Most of these were small in volume and were used in projection lamp applications where the lamp life was sacrificed for high light intensity.

The high operating pressures led to concern about the strength of the quartz and the ultimate safety of the lamp. One of the first high pressure lamps made was General Electric's ELB design, which contained 13 atmospheres of gas. The lamp was intended for use in a projector, so the designers made certain that if the lamp were to fail during operation, any ensuing damage would be confined to the equipment and not to people.

There would be no problem in filling lamp envelopes made of quartz to high fill pressures if the moly seal ribbon was not used. When the moly ribbon is added, the ultimate pressure containment depends upon proper heat control during sealing, and also in obtaining the proper configuration of the seal. Seals with sharp entrant angles in the pressed quartz as it meets the ribbon are not very strong (see Figure 4-3).

Because of the advantages of using high fill pressures, it became necessary to devise suitable means for measuring those pressures in completed lamps. One technique is to gradually fill a group of unlighted lamps with increasing increments of compressed air, subjecting them to a strong internal pressure to make them break. When half the group fails, the pressure applied is an indication of what that lamp design can withstand.

To ensure the safe operation of high pressure lamps, some types were checked 100% by pre-filling them at a pressure slightly exceeding the highest expected operating pressure.

Hard Glass Bulbs

As halogen lamps began to be used in general lighting applications, it became judicious and competitive to change from expensive quartz to some alternate envelope material. Hard borosilicate glass was chosen. Such envelopes were first used in automotive headlamps and later on in PAR lamps.

Hard glass is not only less expensive but its envelope can withstand a pressure of 4-5 atmospheres. Hard glass is more susceptible to thermal shock than quartz, so halogen lamps made with hard glass are generally limited to use in applications where the lamp is protected by either a heavy glass or plastic enclosure, if perchance, it may shatter at end of its life.

With hard glass, the feed-throughs in the seal can be rods, instead of ribbons of molybdenum, which simplifies automation and provides lower cost.

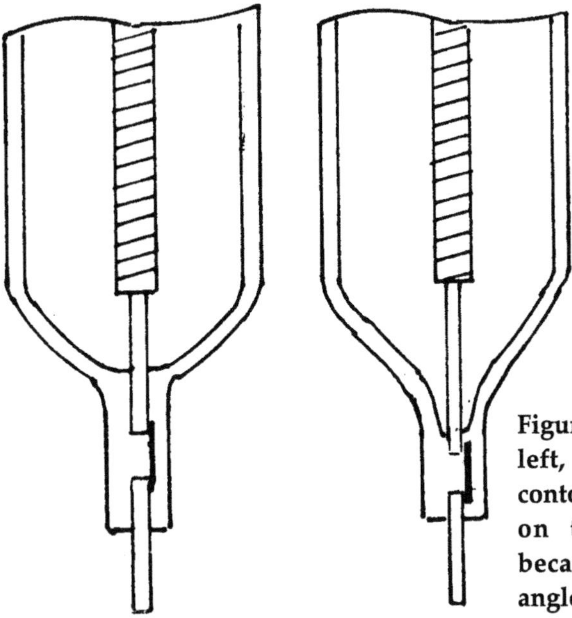

Figure 4-3. The seal on the left, with a smooth entry contour, is strong. The seal on the right is weak because of a sharp entry angle.

Infrared Reflective Coatings

As in all incandescent lamps, the ultimate efficacy of halogen lamps is limited to the melting point of the tungsten filament. In the 1970s, the Duro-Test Company patented a revolutionary, precision-formed, spherical incandescent bulb with an infrared reflecting coating[11], but was not able to produce a commercial product with that design.

In the early 1980s, General Electric introduced a line of tubular halogen lamps that used a comparable type of infrared coating. The outer surface of the tubular lamp was coated with a multilayered interference filter that transmitted the visible light while reflecting the infrared radiation back onto the filament. Tubular lamps, having a concentric filament mounted inside, lend themselves to this principle.

The General Electric tubular lamps did become popular, and gains in lumen efficacy of up to 30% were reported. Reflecting the heat inwardly, allowed the use of heavier coils which could operate at the same temperature, producing an increase in lamp life.

With this method, they were also able to replace a 500W linear flood lamp with one of 350 watts, without any loss in light output. The concept led to the development of small infrared lamps that could be mounted inside of PAR envelopes. These lamps have an efficacy of 20% more than comparable halogen lamps without the infrared film.

Dichroic Coatings

Earlier dichroic coatings worked just the opposite than the IR coating because they transmitted the infrared radiation and reflected the visible light. They were first used on projection lamp reflectors, which absorbed the heat from the lamp and directed only the light forward, thereby preventing the film from melting. The coating consisted, basically, of three inference color filter layers, one for "red," one for "green" and one for "blue." The entire color spectrum is emitted by the lamp filament; therefore, it is white light that is reflected by the combined layers.

The dichroic layers can be deposited by sputtering or by chemical deposition. Their effective life depends on the amount of heat and exposure time to which they are subjected. The coating characteristics are controlled by the thickness of the alternating layers of silicon dioxide and some additional material, such as magnesium fluoride, titanium dioxide and others. Because of their differences in coefficient of expansion, different combinations may be required, depending on whether the coating is to be deposited on quartz, hard glass or some other material.

To make the coating work, requires the precise location of the filament during its entire life, which can be more than 2000 hours. To obtain filaments which have the proper non-sag properties to do this, required making extensive improvements in coil recrystallization, grain structure formation and tensile strength uniformity.

End Seal Characteristics

There is no known metal that has a coefficient of expansion low enough to match that of quartz. Molybdenum comes close, and for sealing, it is rolled into ribbon form with the edges etched to a sharp wedge point. When the quartz is pressed around the ribbon, a hermetic seal is formed. The expansion rates change even more at elevated temperatures and the moly drives into the quartz as a wedge, increasing the tightness of the seal. The thin edge of the ribbon is not strong enough to force the quartz seal to open.

The lamps must operate at a high temperature in order to maintain the halogen cycle. If the seal temperature rises too high during opera-

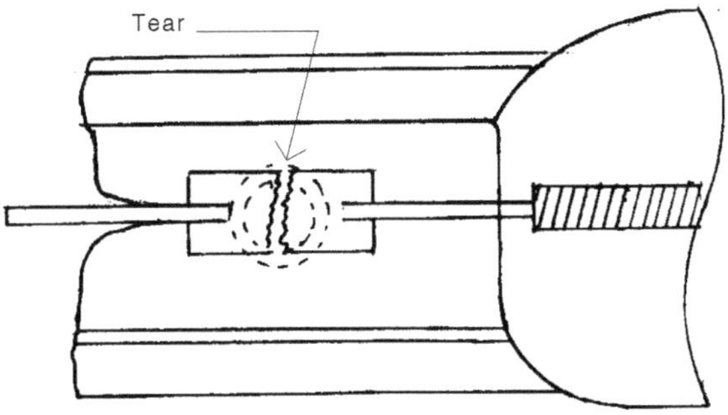

Figure 4-4. Torn molybdenum ribbon due to overheating from excessive seal temperature.

tion, however, the embedded moly ribbon and end rod will oxidize when the air temperature exceeds 350°C which will cause the seal to crack or the ribbon to tear. Figure 4-4 shows what a burned out ribbon looks like.

In those applications where much higher temperatures are required, the capillary path between the quartz and the moly is sealed with a molten glass or frit such as lead borate.

Applications

All tungsten halogen lamps have the following desirable characteristics:

1. They provide high wattage and high light output in a relatively small volume.
2. They provide longer life without sacrificing lamp brightness or efficacy.
3. They provide almost constant light output throughout life.
4. Because of their small size, they permit the use of smaller, less expensive fixture and lenses.

Halogen lamps have found broad application in two different areas of general lighting usage:

1. As a long, linear light source—around which are designed unique and highly efficient luminaires which provide a beam of broad, bright light for use in floodlighting, store and building lighting, and stage and studio lighting.

2. As a small, compact and brilliant light source of high optical precision. Tungsten halogen lamps have replaced many of the applications long-served by incandescent lamps, and because of their ruggedness, have even created some new ones. Halogen lamps are used extensively in airfield lighting, photographic work, automotive lighting and in fiber optic devices.

The bright long-lasting light provided by all of these halogen light sources has made them an integral part of lighting design within the short span of two decades.

Linear Lamps

The linear lamps are used in the following areas:

Floodlighting

These tubular lamps, with their accurately positioned filaments, have found use in many floodlighting applications. The lamps are often mounted in trough-like reflectors which produce narrow fanlike beams having good, sharp cut-off patterns. These patterns are desirable because they produce a distinct cutoff which shields passersby from glare. Figure 4-5 shows a suitable beam pattern that can be produced from a well-designed fixture.

The fixtures are built with either parabolic or elliptical reflectors to focus the light beam. The units are broadly used in sports arenas, outdoor automobile sales lots, tennis courts and parking lots.

Store Lighting

In stores and showrooms, the lamps are used for merchandise display lighting. They provide excellent color rendition because they radiate a complete and continuous visible spectrum. In stores, most halogen lamps are the PAR bulb variety and are used in downlighting fixtures, including the IR PAR, the MR11 and the MR16 reflector types.

Figure 4-6 shows a typical dichroic coated 50W MR16 lamp.

Stage and Studio Lighting

Photographic and television studios use halogen sources because of their good color rendition and high color temperature. Early television cameras required sources of 3200K to 3400K to match the film used. Today, the cameras and film are more versatile, but halogen lamps are

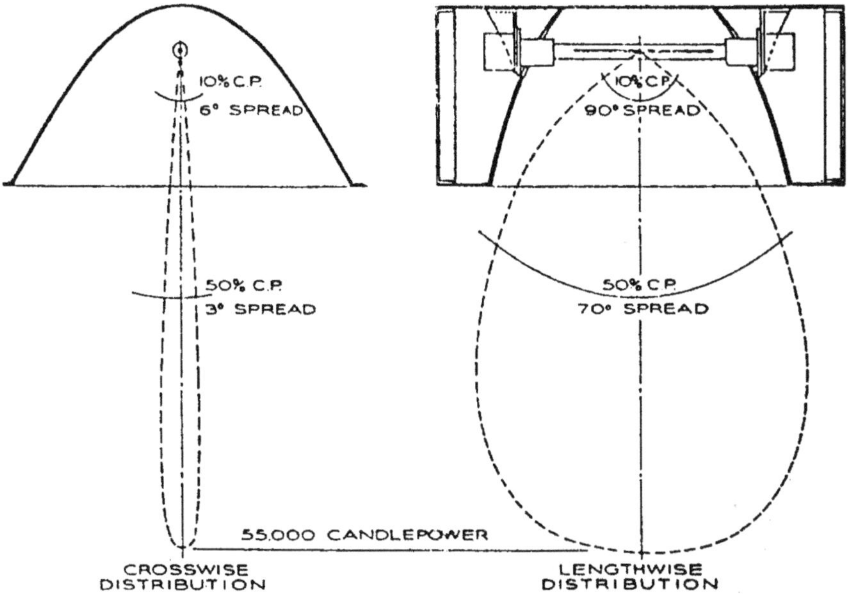

Figure 4-5. Carefully controlled beam pattern in a floodlighting fixture, made possible by the unique and compact linear design of the halogen lamp, side view (left) and frontal view (right). (From Allen & Paugh, 1959.) Courtesy of the Illuminating Engineering Society of North America.

still being used.

A popular and related area where halogen lamps are used in great abundance is lighting stages and concert areas where intense light is needed. Studio and stage lamps are generally made in high wattages, from 1000-5000W, because of the high brightness required and because they must often be mounted in lofts or on high masts, at considerable distances from the subjects. Figure 4-7 shows a typical halogen stage light.

Single- and Double-ended Halogen Lamps

The single- and double-ended halogen lamps are used in these areas:

Airfield Lighting

High-intensity compact halogen lamps are used in many airport applications. Many of them have specially designed filaments to permit

Figure 4-6. A 50W MR16 halogen lamp with a dichroic coating applied to the back portion to transmit the radiated heat through the reflector. Courtesy: Osram Sylvania.

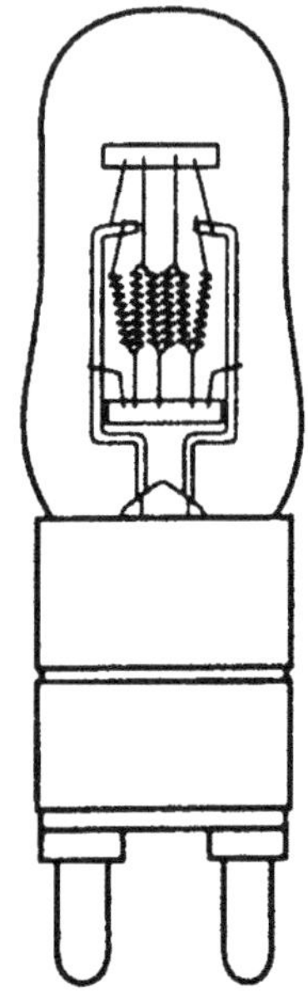

Figure 4-7. A T20 5000W mogul bipost halogen lamp used in studio lighting. Courtesy: GE Lighting.

operation in series on constant current circuits. Their brilliance and long life make them ideally suited for this use. For example, 45W or 200W, 6.6A lamps are made to fit inside of small, rugged and flat fixtures that are placed into shallow holes drilled into the concrete of the runways (see Figure 4-8). A string of these lights will appear as a straight line to the pilot to show him the proper approach for landing. He can also follow them to the taxi area. Other sizes are also used in airport guidance systems and beacons.

Photographic Lights

High-wattage halogen lamps housed in portable reflector units have been designed for still and movie photography. These are generally made in 625W and 800W sizes. Others are made in wattages of 1000W and above, for indoor and outdoor scenes, especially for "fill-in" light-

Figure 4-8. A "pancake" fixture which can be embedded in a concrete airfield runway. It houses a rugged 200W 6.6A halogen lamp.

ing effects. These larger units can also be fitted with spreader lenses to create various beam patterns, or filters to produce desired changes in color.

They are also used in movie projectors, although these are now being replaced by videotapes. With the growing popularity of camcorders, halogen lamps can be used to provide auxiliary lighting.

There was a time when these lamps were used for microfilm and microfiche readers, but that equipment is quickly being replaced by computers and CD-ROM disks.

Medical Instruments

Compact halogen lamps have found many applications in medicine, where they are used in operating rooms and in medical equipment. Some serve as the light source in fiber optical devices, which are used to examine various parts of the human body that cannot be observed by any other means. Others provide the ultraviolet curing light in dental applications. In fact, the light over your dentist's chair is probably a halogen lamp.

Automotive Lamps

Compact halogen lamps now dominate the automotive headlight systems. They are usually embodied in as single-ended units, mounted on their metal or plastic base and inserted into hollow metal reflectors. They are also mounted inside of PAR-Type and REC-Type (Rectangular) sealed beam units. Figure 4-9 shows some of the common halogen lamps now used in automotive headlights.

Their precise filament location permits the lamps to be positioned at the focal point of their reflector housing. This results in brighter and controlled illumination, to make driving safer at higher speeds during the night.

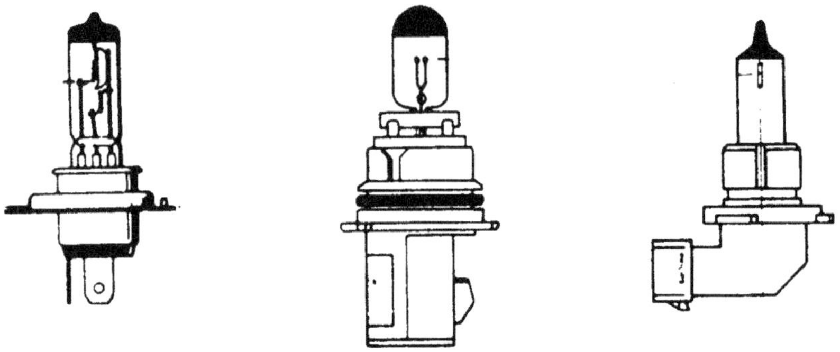

Figure 4-9. Three types of halogen lamps currently used in automobile headlights, 9003 lamp (left), 9004 lamp (middle) and 9006 lamp (right). Courtesy: Philips Lighting.

Accent Lighting

As a bromine light source, accent lights are housed in their own dichroic-coated reflector envelopes. They are used in store showcases, down-lighting and in portable lanterns. Their acceptance and growth has been remarkable. The most popular types are the MR16s that are made in the 20W, 50W and 70W sizes.

References

1 Scribner, E.A., "Electric Incandescent Lamp," U.S. Patent 254,780, 1882.
2 Anthony, W.A., "On the Effect of Heavy Gases in the Chamber of an Incandescent Lamp," *Electricity*, March 1894.
3 Van Liempt, J.A.M., "Electric Incandescent Lamp," U.S. Patent 1,925,857, 1933.
4 Moore, J.A. and Jolly, C.M., "The Quartz-Iodine Tungsten Lamp," *GEC Journal*, 1962.
5 Fridrich, E.G. and Wiley, E.H., "Electric Incandescent Lamp," U.S. Patent 2,883,571, 1959.
6 Zubler, E.G. and Mosby, F.A., "An Iodine Incandescent Lamp with Virtually 100 Percent Lumen Maintenance," *Illuminating*

Engineering, 1959.

7 Schilling, W., "Gas-Filled Electric Low-Voltage Incandescent Lamp," German Patent 1,183,596, 1965.

8 Van Tijen, J.W., "Iodine Incandescent Lamps: Principle," *Philips Technical Review*, 1961/62.

9 English, J.F. and Zubler, E.G., "Iodine Cycle Incandescent Electric Lamp," U.S. Patent 3,240,975, 1966.

10 Bonazoli, R.P. and Scoledge, R.F., "Incandescent Lamp Containing Iodine, Oxygen and Inert Gas," U.S. Patent 3,242,372, 1966.

11 Brett, J., Fontana, R., et al, "Development of High Energy-Conserving Incandescent Lamps," *Journal of the Illuminating Engineering Society*, 1982.

Chapter 5

Lamp Phosphors

by Jacob Van Broekhoven

Introduction

A phosphor is any substance capable of absorbing energy and re-emitting it in the form of visible light. Phosphors make the fluorescent lamp work. A fluorescent lamp is a very efficient generator of ultraviolet energy, mostly at about 254 nanometers (nm), and some at 185 nm. Fluorescent lamp phosphors absorb this ultraviolet radiation and convert it to visible light. In the lamp industry, the term "light" is expanded to include the near ultraviolet and infrared regions. Phosphor, in the form of a fine powder, is applied as a uniform coating onto the inside surface of the lamp tube, resulting in a very efficient light source.

Phosphor Preparation

A typical phosphor preparation procedure includes these basic steps:

1. Weigh the components.
2. Mix thoroughly. Coprecipitation is sometimes used.
3. Heat to react and form the desired crystal structure.
4. Reduce to a powder. In some cases additional heating and reduction to a powder is required.
5. With some phosphors, a post treatment is beneficial.

Phosphors are typically crystalline, and when fired are often somewhat sintered. They are typically milled to a desired particle size. The milling process is detrimental to the quality of phosphors, but the effect varies greatly with different phosphor systems. The phosphor must be fired at a temperature sufficiently high to develop a good crystalline structure, but must also be low enough so the fired cake can be reduced to a powder with minimum damage. The sintering and milling response varies greatly among the various phosphor systems. The initial reduction to a powder is usually accomplished with special high quality hammer mills, often followed by additional controlled milling in ball mills or other competitive devices.

Phosphor Efficiency

Each quanta of ultraviolet radiation absorbed by the phosphor can theoretically emit one quanta of visible light. The ratio of quanta emitted to quanta absorbed is known as the quantum efficiency. Useful phosphors have quantum efficiencies in excess of 90%. There is always an energy loss because the energy of a quanta of the emitted longer wavelength radiation (light) is less than the energy of a quanta of the absorbed ultraviolet.

Qualifications of An Effective Phosphor

A phosphor must meet a number of qualifications in order to be useful for lamp application. It must:

1. Efficiently absorb ultraviolet radiation.

2. Convert and emit the absorbed ultraviolet in a visible region of the electromagnetic spectrum.

3. Be in the form of a fine powder so as to result in a smooth, aesthetically pleasing coating.

4. Be sufficiently stable to withstand the application process.

5. Withstand, with minimum degradation, the rigorous environment of the operating lamp. This includes exposure to mercury ions and intense ultraviolet radiation. Particularly damaging is the 185 nm mercury radiation.

6. For HID lamp applications, the phosphor must perform satisfactorily at the high operating temperature of about 350°C, but is not subject to mercury ion bombardment.

Method of Application

The method of application requires that, from a practical standpoint, the phosphor used for coating must be a powder. Phosphor powder is made into a paint that is applied to the top inside of the vertical lamp tube, and drained in a manner so as to form a very uniform coating. The phosphor coating is dried and heated to near the melting point of the glass to burn off the organic component of the paint. Binders, usually borates or very fine aluminum oxide, may be added to the paint to improve the adhesive properties of the phosphor and thereby prevent chipped coating.

Phosphor Particle Size

In application, the maximum allowable individual phosphor particle size is about 35 microns, since a larger size produces a grainy coating. From an economic as well as adhesion characteristic standpoint, the average particle size should be as small as possible, consistent with acceptable lamp efficiency. In practice, the average particle size is generally below 10 microns.

The effect of particle size requires some explanation. The optimum coating thickness is a function of the optical characteristics as well the total phosphor mass, just as the covering power of a pigment in the paint industry is a function of the particle size and total quantity of pigment. The smaller the particle size, the greater the covering power and the lower the total weight required for a given covering power.

Consider a fluorescent lamp with no phosphor coating. The ultraviolet emission is absorbed by the glass and there is no light output except the small amount of blue from the mercury arc. As phosphor is gradually applied, more ultraviolet is absorbed by the phosphor and less by the glass. At some point the output levels off, and then begins to decline, as the additional phosphor no longer absorbs additional ultraviolet, but does absorb some visible light since it is not completely transparent. The optimum phosphor coating weight decreases as the particle size decreases. At the same time, the lumen output and maintenance of most phosphors decrease with decreasing particle size, the effect differing widely from phosphor to phosphor. It is generally advantageous to minimize the quantity of particles below one or two microns in order to maximize lamp efficiency.

History

Phosphors were known as far back as 1602 AD when Vingenzio Cascariolo of Bologna, Italy, heated a mineral, a natural sulfide of barium, while experimenting with alchemy and produced a product that glowed strongly in the dark after exposure to bright light. In 1866 Theodore Sidot prepared a zinc sulfide phosphor by heating zinc oxide in a stream of hydrogen sulfide. This was the forerunner of our present day cathode ray phosphors.

The First Useful Application

Possibly the first useful application of a phosphor was by Thomas A. Edison in a patent applied for in 1896, granted in 1907, covering a short vacuum tube coated on the inside with blue emitting calcium tungstate phosphor. The excitation was actually cathode ray, but it was a forerunner of the fluorescent lamp. The first excellent working model of our present day fluorescent lamp was produced by W.S. Andrews in 1912. It employed cold cathodes and had low efficiency.

The First Commercial Fluorescent Lamp

In 1935, General Electric (GE) initiated a project to commercialize the fluorescent lamp based upon the early 1920s work of Dr. Hull and Dr. Langmuir. The effort was aided by an inventory of known phosphors from which to choose. They were mostly man-made copies of naturally occurring fluorescent minerals that were more or less optimized in the laboratory. In 1938, GE announced the fluorescent lamp as an addition to their regular line of lamps after incurring a development cost of only $107,000, a modest sum even for that time. Progress was put on hold with the advent of hostilities in Europe and the onset of World War II.

With the cessation of hostilities, progress began in earnest. A vital factor in continued progress was the availability of suitable phosphors. These phosphors had to meet a rather stringent set of criteria as previously discussed. Basically, they had to be inorganic to survive the application technique and stable enough to withstand the harsh environment inside the lamp.

A Review of Early Inorganic Phosphors

Inorganic phosphors are minerals containing small amounts of impurities known as activators. Many early phosphors were simply optimized copies of fluorescent minerals found in nature. Examples are zinc sulfide, usually activated with copper or silver, extensively used in cathode ray tubes, and green-emitting manganese activated zinc silicate (commonly known as willemite).

The typical phosphor is produced by thoroughly mixing a number of components containing the required elements, and firing them at the temperature required to initiate a reaction that will result in the desired crystal structure. In some cases, the components are coprecipitated, rather than mixed, and then fired. The reaction is essentially solid state, but often involves a gaseous atmosphere, either self-produced or introduced. The required atmosphere may be oxidizing, neutral or reducing. Many phosphors are a synthesis of naturally occurring minerals, but many others are the product of scientific inquiry.

The fluorescent lamps developed in the late 1930s and most of the 1940s for general lighting contained orange-emitting manganese activated zinc beryllium silicate phosphor, blended with some blue phosphor to obtain white light. (Of course, white light is a matter of definition and in practice consisted of a range of black body color temperatures from about 3000K to 6500K at approximately the black body locus.) The lamp, with an efficacy of 45 lumens per watt (LPW), was a vast improvement over the 12-15 LPW delivered by the incandescent lamp. Fluorescent lamp life was about 2,000 hours compared to 750 hours for the incandescent lamp (present-day fluorescent lamps have an average service life of 20,000 hours).

Shortly after World War II ended, a most disturbing revelation presented itself. Beryllium was toxic! It was a strange toxicity in that it was not toxic to everyone. A secretary working for someone with high exposure to beryllium could be fatally poisoned, while the highly exposed person would not be affected. Several fatalities occurred, but there are people alive and well today who had high everyday exposure, for several years. This toxicity was not immediately recognized because the toxic results took a relatively long period of time to manifest themselves, and only a small percentage of people were susceptible. It became obvious that Zn Be silicate use had to be discontinued. A period of extensive phosphor research and development followed.

Some of the phosphors that were available before 1950 include:

Name	*Formula*	*Activator*	*Emission Color*
Zinc Silicate	Zn_2SiO_4	Mn	Green
Zinc Beryllium Silicate	$(Zn,Be)_2SiO_4$	Mn	Orange
Calcium Silicate	$CaO{:}SiO_2$	Pb, Mn	Salmon
Cadmium Borate	$CdO{:}B_2O_3$	Mn	Pink
Calcium Tungstate	$CaWO_4$	Pb	Blue
Magnesium Tungstate	$MgWO_4$	Self	Blue

These available phosphors could be blended to produce white light and thereby eliminate the toxic zinc beryllium silicate, but with a penalty in cost and lumen efficiency. But fortunately, McKeag and Ranby, G.E.C. England[1], developed a new phosphor that appeared to be a perfect substitute. It has an apatite crystal structure with the basic formula of $Ca_{10}(PO_4)_6(F,Cl)_2{:}Sb,Mn$. The name calcium halophosphate was given to the system.

The beryllium toxicity forced the industry to discontinue zinc beryllium silicate and convert to the calcium halophosphate system as rapidly as possible. Calcium halophosphate almost immediately became the dominant phosphor in the fluorescent lamp industry and will probably continue to be for many more years. The rare earth phosphors are becoming increasingly important and will also be discussed in this chapter.

Calcium Halophosphate Phosphors

Calcium halophosphate phosphors are modified versions of the crystal system known as apatite, with the formula $[Ca_3(PO_4)_2]_3{\cdot}CaF_2$. This phosphor is prepared by replacing a small amount of the calcium (Ca) with antimony (Sb), as the primary activator, and manganese (Mn), as the secondary activator. Without Mn, the emission is blue. Mn addition gradually changes the color from deep blue to orange. The phosphor has two broad emission bands, blue from the Sb and orange from the Mn. Increasing Mn suppresses the Sb band. Replacing some of the fluorine (F) with chlorine (Cl) shifts the long wavelength end of both bands in the red direction.

The formula for calcium halophosphate phosphor can be expressed as:

$$[Ca_3(PO_4)_2]_3.Ca(F,Cl)_2{:}Sb,Mn$$

The spectral power distribution (SPD) curve for a standard cool white phosphor is shown above in Figure 5-1. The effect of the two peaks in a real phosphor is evident.

With this one phosphor system, it was possible, by controlling composition, to produce highly efficient phosphors with colors ranging from about 7000K to 2700K, approximating the black body locus of the ICI color diagram. In addition, the required components were readily available at low cost.

Components

When the sudden transition to calcium halophosphates began, the components were purchased from existing vendors. Various chemical compounds were contenders as long as they contained the required calcium, phosphate, manganese, antimony, fluorine and chlorine. As development work progressed, the materials listed below were found to be best from the standpoint of quality, convenience, cost and availability:

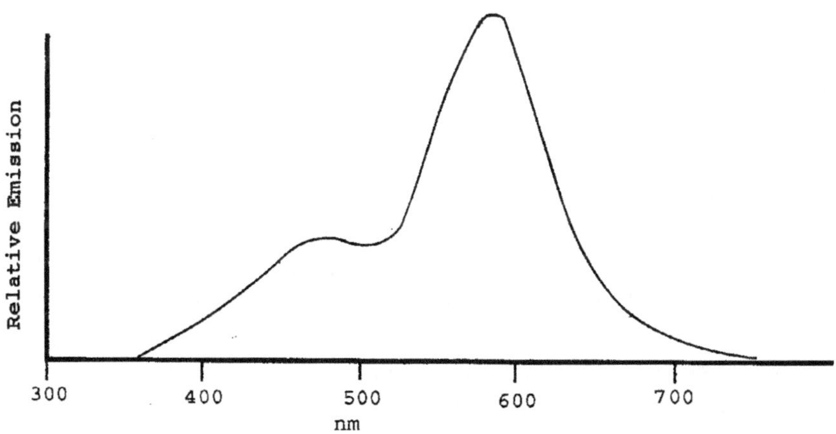

Figure 5-1. Cool white calcium halophosphate emission.

$CaHPO_4$	Calcium Phosphate Dibasic
$CaCO_3$	Calcium Carbonate
CaF_2	Calcium Fluoride
NH_4Cl	Ammonium Chloride (Not used in deep blue)
Sb_2O_3	Antimony Trioxide
$MnCO_3$	Manganous Carbonate (Not used in blue or deep blue)

At the beginning, Mallinckrodt Chemical Co. made a decision to become the major component supplier for the phosphor industry. At first, it was successful in this effort, but as the industry advanced, the phosphor manufacturers desired complete control of the quality and cost. Sylvania and General Electric began to produce their own dibasic calcium phosphate ($CaHPO_4$), calcium carbonate ($CaCO_3$) and calcium fluoride (CaF_2) in the 1960s. Westinghouse, now Philips Lighting, continued using Mallinkrodt produced materials until about 1980, when it began making its own $CaHPO_4$ and $CaCO_3$.

Calcium Phosphate Dibasic

The $CaHPO_4$ is a very important ingredient because it is the largest component as well as a major factor in controlling the particle size of the finished phosphor. A process that produced a good high-purity uniform product was soon developed. Diammonium phosphate and calcium chloride solutions were reacted to precipitate dibasic calcium phosphate dihydrate ($CaHPO_4.2H_20$), which was then dehydrated by heating to near the boiling point of water while still in 'a slurry. The dehydration was accompanied by a recrystallization that excluded most of the impurities, resulting in a very pure product suitable for phosphor manufacture. Accurate control of precipitation and dehydrating conditions was essential in maintaining purity, particle size and particle size distribution. It was desirable to maintain as narrow a particle size distribution as possible

After the dehydration, the $CaHPO_4$ was filtered off and, at first, the effluent—containing mostly ammonium chloride—was discharged. Because of environmental considerations, the process was modified so as to recycle the ammonium chloride waste. It was treated with calcium oxide and heated to drive off the ammonia as a gas, which was absorbed in phosphoric acid to yield diammonium phosphate for reuse. The calcium chloride solution remaining after the ammonia recovery was purified and used to produce additional $CaHPO_4$.

Calcium Carbonate

The calcium carbonate was manufactured by reacting ammonium carbonate and calcium chloride solutions in a continuous reaction at room temperature to precipitate vaterite as spherical particles. After filtering, the vaterite was re-slurried and heated to 90-95°C, at which point it recrystallized to the more stable and crystalline form known as calcite, and at the same time excluded most of the impurities.

Other Components

A pure, fine, crystalline type of calcium fluoride is now used that is manufactured especially for phosphor manufacture. It is produced by Osram Sylvania and General Electric. Sylvania acts as supplier for other phosphor manufacturers. Nichia, Japan, is a competitive world supplier of calcium fluoride and other phosphor chemicals and phosphors.

The remaining ingredients, antimony trioxide, manganous carbonate, and ammonium chloride are purchased from quality commercial producers.

Formulation

The formulation was optimized experimentally. The optimum formula differed somewhat from company to company because of variations in manufacturing techniques, but the differences were not great.

The ammonium chlorine and antimony trioxide requirement was always in excess of theoretical. The ammonium chloride reacted to form calcium chloride and antimony trichloride as intermediates. Some of the antimony trichloride volatilized and was lost, the volatiles forming a combination of reactive and protective atmospheres. The firing procedure was an important factor influencing the required excess. Too high an excess of antimony forms inert, acid-insoluble, calcium antimonate $CaSb_2O_6$ in the phosphor crystals. An acceptable phosphor was found to retain about 0.5-0.7% antimony by weight.

The color was controlled by adjusting the chlorine/fluorine ratio and the manganese content.

The $(Ca + Mn + Sb)/PO_4$ (M/P)ratio is theoretically 10:6. But experimental data required a ratio somewhere between 9.6:6 and 9.9:6. When the ratio was increased above these figures, some manganese oxidized to higher valence states and the fired cake became pink—and in extreme cases, black. The quality degrades precipitously from too high an M/P ratio, but then slowly as the ratio decreases.

Working formulations for several colors are shown below:

		Moles	
Component	*Blue*	*Cool White*	*Warm White*
$CaHPO_4$	6.0	6.0	6.0
$CaCO_3$	2.65	2.582	2.486
CaF_2	0.90	0.90	0.85
$MnCO_3$	—	0.178	0.364
NH_4Cl	—	0.36	0.36
Sb_2O_3	0.030	0.064	0.06
Ca+Mn/6P	9.55	9.66	9.70

As the manufacturing process evolved, it was discovered that by replacing some of the calcium with cadmium, in the amount of about 0.5-1.0% by weight of the phosphor, a lumen improvement of 2-3% was realized. This became an industry standard until it was realized that cadmium posed an environmental threat, which was strongly evidenced when cadmium compounds, used extensively as a fungicide for lawns and golf courses, eventually entered the food chain. The use of cadmium in phosphors was discontinued throughout the United States about 1985, several years after the rest of the world did.

The industry explored the periodic table for a cadmium substitute, but with essentially no success. The only additive found to date that produced an effective improvement was gallium, which was rejected because of its excessive cost.

Formulation Calculation

Before formulating a raw mix, it was necessary to assay the components in order to accurately add the individual elements. But even after the industry progressed to the point where the components were manufactured by consistent standardized methods, and manufacturing procedures seemed to be well-controlled, there continued to be unexplained variations in color and quality. In an effort to improve product uniformity, a critical study was made of the analytical procedures used to assay the components, and the standard deviations of the results. The standard deviations were found to be considerably greater than required to achieve good manufacturing control. In addition, microscopic examination of the $CaHPO_4$ indicated that the crystals were highly transparent and very well formed. From these observations it was concluded that the calcium-to-phosphorus ratio must be very consistent for the crystals to be so well formed. If this were indeed the case, it would be reasonable to assume that the calcium and phosphate in the $CaHPO_4$ was stoichiometric and essentially constant, and the only assay that would be of value was the moisture content. In practice, it was more convenient to determine the "loss on ignition" (LOI) and obtain the equivalent calcium pyrophosphate $Ca_2P_2O_7$ value. The same argument was applied to the calcium carbonate and the calcium fluoride; that only the moisture content needs be determined and used in formulation calculation. When this procedure was initiated, product uniformity—both in lumen output and color—improved markedly. This consistency was a great aid in process optimization. Manganous carbonate is not a well-defined compound, and therefore continues to require a manganese assay.

There are some very able people in the phosphor industry who disagree with the author in assuming that the components are constantly stochiometric, and only moisture content is of value in calculating formulations. While they are theoretically correct, the experience at Westinghouse found this alternate approach to be advantageous. Perhaps the precision of advanced analytical methods will eventually detect small variations in the composition of the well formed crystals, but the uniformity of product when using this technique is so good it would seem difficult to justify the added effort.

Mixing

In the early mixing process, components were loaded into a ball mill and milled for a period of time sufficient to accomplish satisfactory mixing. Unloading was accomplished by putting a screen in place of the loading port cover and running the mill until all of the mixed powder fell into a collector. This was labor- as well capital-intensive, requiring a bank of ball mills and large floor area per unit of production. A better method was needed.

Various methods of mixing were investigated. An extremely thorough degree of mixing was required because the reaction was largely solid state. Double conical mixers—used in other industries for many years—were tried, but found to be deficient as to the degree of mixing.

However, a double conical mixer containing a high-speed mixing device was then installed. It was patented and manufactured by the General Machine Co. of Newark, NJ. Numerous mixing disk configurations from double-shearing to single breaker were tested. Eventually a single high-speed serrated breaker disk was found to work best, and large-size models were produced. This type of mixer is still used, with improvements, of course. A "V" type blender with a breaker wheel is manufactured by Patterson Kelly and is the mixer of choice of several manufacturers. Both types are quite competitive and produce similar results. Large-size mixers with automatic weighing, loading and unloading now provide fast, efficient and economical mixing of phosphor components.

Furnaces and Firing

Zinc beryllium silicate phosphor was fired in open trays in air atmosphere box furnaces heated with silicon carbide heating elements. Initial production of the new calcium halophosphate phosphor was fired in these same furnaces. A neutral atmosphere to keep the Mn^{2+} from oxidizing to the 3+ or 4+ valance state was now required. In an oxidizing atmosphere the product is a pink to black color, obviously of low brightness. A neutral atmosphere was obtained by firing the raw mix in a set of silica containers that consisted of a tray and overlapping cover combination that was inverted prior to firing so as to produce a powder seal. The gases emitted during the decomposition of the raw materials provided the required neutral atmosphere. The phosphor forming the seal was exposed to air and became pink. This material was unusable and was removed and discarded. In order to improve the efficacy, the fired cake was reduced to a powder and again fired, typically with the addition of a small amount of antimony or a chloride or (sometimes) both.

As production requirements increased, the batch furnaces became undesirable because of high labor costs, high power costs and excessive floor space requirements. The first automated furnaces used were stoke-type electric furnaces in which the batch furnace type silica trays, with overlapping inverted covers and inverted to form a powder seal, were put onto silicon carbide slabs and automatically pushed through the furnace at a rate determined by the required firing time. The first furnaces were manufactured by Harper Furnace Co. using a rather standard design. These were air atmosphere furnaces with a door at the entrance end and an open exit end. The stoke cycle was periodic—that is, the door was opened, a tray was pushed into the furnace and the door then closed until time for the next stoke.

The furnace configuration from front to back consisted of an idle zone, a preheat zone, a full temperature zone and a cooling zone. The exhaust gases were vented in the idle zone so as to preheat the incoming trays. The preheat zone heated the trays at a controlled rate. The full-heat zone completed the reaction, and the cooling zone cooled the trays as well as preheated the incoming air. It was important to control the heating and cooling cycle to avoid excessive thermal shock to the silicon carbide pusher tiles that carried the trays through the furnace. Excessive thermal shock would break the tiles and jam the furnace. Since these furnaces were about 60 ft. long, a jam-up would cause a major production interruption. The silica trays, unlike the tiles, could tolerate severe thermal shock, but would eventually break, sometimes catastrophically, due to attack by the fluorine evolved during the reaction, and compounded by the fact that the required firing temperature was somewhat above the recommended maximum for silica.

These furnaces were significantly more efficient than the batch furnaces, but were subject to a build-up caused by the condensation of materials evolved from the reaction, eventually narrowing the tunnel until the trays could no longer pass through. These episodes occurred at intervals of only several months. Furnace shutdowns and partial furnace rebuilding were quite frequent. As production requirements grew, additional furnaces were added, the later ones with two parallel tunnels.

At first, the tray and cover configuration was the same as that used in the batch furnaces—that is, a tray and overlapping cover that was inverted to produce a powder seal. The disadvantage of this was that the phosphor making the seal was exposed to air and was pink, unsatisfactory for use. This material was discarded at first, but as time went on attempts were made to reprocess it, with varying degrees of success, influenced by changes made in the continuous effort to improve the product. Eventually, the air flow through the furnaces was decreased and the formulation was adjusted so that the trays did not require inverting, and there was no pink material to recover. These adjustments reduced, but did not eliminate the build-up problem. Other improvements in materials and formulation eventually eliminated the need to double-fire, resulting in significant savings in labor, power, tray and capital costs.

Eventually, the industry converted to a process of firing in a neutral atmosphere of nitrogen. A new type of furnace was now required, supplied by American Electric Furnace Co., first used by Sylvania. The basic configuration was similar to the Harper furnaces, except that there were double doors at the entrance and exit with a purge chamber in-between. Nitrogen was introduced mostly at the front and rear of the furnaces and exhausted at a choice of several built-in vent positions. This furnace improved phosphor quality, and also reduced costs by not requiring covered trays. However, the elimination of covers had a dual effect. On the one hand, greater uniformity from tray to tray was achieved because it eliminated the variation in cover integrity. On the other hand, since the trays were uncovered and the atmosphere varied throughout the furnace, control of gas flow and furnace exhaust became much more critical.

The build-up problems encountered in the Harper furnace were also encountered in the American Electric furnaces. Initially, the build-up would reach a critical level, necessitating a shut-down in 6-12 weeks. Eventually, by judicially adjusting vent position, nitrogen flow and formulation to reduce volatile emission, the rate of build-up was reduced to the point where the period between shut-downs because of build-up were extended to 4-8 years.

In retrospect, the solution was quite simple. The nitrogen was introduced into the furnace mostly at the front and the back so that the gas flow was in from both ends and out the vent. When the vent was at the front low temperature area, the gas flow from the rear traveled from the cooling section, through the hot zone, into the preheat zone and out the stack. The gas flow carried the phosphor volatiles, some of which condensed when entering the cooler sections of the furnace, resulting in a build-up that eventually clogged the furnace tunnel. By changing the venting to the first part of the full temperature zone, the volatiles were discharged at the highest temperature and did not have the opportunity to condense. This vent change exposed the phosphor in the open trays to a new mode of atmospheric exposure that upset the previously optimized conditions. With proper reformulation and gas flow adjustments a new set of optimum conditions were obtained, with essentially no build up in the furnace. The operating parameters of the furnace become more critical when venting from the full temperature area.

At the present time, the industry, at least at Osram Sylvania and

Philips Lighting (having purchased Westinghouse Lamp Division), are using Harper furnaces designed for atmosphere firing.

Firing Container Materials

The firing container cost is an important factor in the total cost of manufacturing an inexpensive phosphor such as the calcium halophosphates. Silica trays were the industry standard for many years. They were readily available in a large range of sizes, and were the same material used in the laboratories to do original research and development. Silica is excellent from the standpoint of thermal shock and resistance to attack by phosphor chemicals. In fact, silica is still the only practical material that can withstand the thermal shock of batch furnace firing where the loaded tray is introduced into, and removed from, the fully heated furnace.

With the introduction of continuous stoke-type phosphor furnaces, the thermal shock requirement was drastically reduced, so that it became feasible to investigate other tray materials that had longer life, but with inferior thermal shock characteristics. The silica trays were used at a calcining temperature approaching 1200°C, while the manufacturers recommended a maximum use temperature of 1100°C. In addition, the fluorine in the phosphor attacked the trays and shortened the life, resulting in a typical life-span of service of between 20 and 30 cycles. Another negative factor was that the end of life was often catastrophic, so that phosphor powder and silica fragments ended up in the furnace, often causing the pusher tiles to jam-up—resulting in furnace downtime at best, and severe furnace damage at worst.

In an effort to reduce tray costs, a large number of ceramic materials were tried, mostly resulting in failures. Alundum was found to be satisfactory from a quality standpoint, but the densely fused alundum available in small laboratory ware was extremely costly to produce in sizes large enough for use in phosphor manufacture. A pressed and sintered version of the alundum tray was available at a competitive price, but it required a thick wall dimension to achieve satisfactory strength, so that the tray, and therefore the furnace, capacity was unacceptably reduced. The thick wall also resulted in a rather heavy tray that presented some handling problems. In addition, metal chlorides were absorbed into the porous tray, and if a used tray were stored in a somewhat high humidity environment the chlorides absorbed moisture to form a hydrate. The expanded hydrate caused the trays to break similar to the effect of ice formation.

While the pressed alundum trays were rejected for calcium halophosphate use, they were used to fire phosphors containing alkali metals, which react with silica.

Eventually, Norton Co. developed a tray made of "Crystar," the trade name of a new material that is essentially pure silicon carbide. The tray walls were quite thin, about 0.25 inches, but with reasonably high physical strength. The thermal shock characteristics were not as good as silica, but with the continuous furnaces it was satisfactory. The maximum recommended temperature is well above that required for calcium halophosphates, and has a practical life 10-50 times that of silica, ample to offset the higher cost. An added advantage was that failure is gradual. It starts with a small crack that grows slowly, permitting the tray to be discarded before total disintegration, which could cause the furnace to jam.

Acid Washing

About 1960, it was discovered that calcium halophosphate phosphor quality was improved simply by washing. Improvement was noted by washing with weak alkali, acid and even just plain water. The greatest improvement was noted with weak solutions of mineral acids, generally in the range of 0.1 to 0.2 normal. Hydrochloric or nitric acid solution was the most convenient to use and was equally effective. Sulfuric acid was also effective, but had the disadvantage of requiring extraordinary precautions to remove all traces of sulfate, since even small amounts were very detrimental to fluorescent lamp performance. Washing improved phosphor performance by 2-3%. Some manufacturers treated the phosphor with an ammonium hydroxide wash after the acid wash.

A number of theories were developed to explain the benefits of acid washing, some of them quite complicated. The benefit is probably the result of removing small amounts of extraneous antimony compounds from the surface of the phosphor crystals. These surface deposits are the result of condensation from the furnace atmosphere during the furnace cooling cycle. It is well known that small amounts of free antimony compounds in a calcium halophosphate lamp is detrimental to lamp maintenance, so that it is not surprising that removing deposited antimony compounds is beneficial. This is an interesting phenomena in that the addition of a small amount of antimony trioxide to some other phosphors, such as cadmium borate and zinc silicate, improves lamp maintenance markedly.

Particle Size Control

In the initial period of phosphor manufacture, the fired cake was milled to a powder and shipped to the lamp plant. There was no attempt to control particle size or particle size distribution. The lamp plant added the phosphor powder and a suitable lacquer to a ball mill, and milled the phosphor until the particle size was reduced sufficiently to result in a smooth lamp coating. The paint was applied to the top of the vertical bulb and allowed to drain and dry in a controlled manner so as to form a uniform coating. If the paint was insufficiently milled, large particles produced unsightly streaks, but excessive milling lowered lamp output. In addition, under-milled or large particle size phosphor did not adhere well, resulting in chipped coating during lamp handling. The optimum coating thickness for a small particle size phosphor was less than for a larger particle size of the same phosphor, thereby introducing an economic factor. In general, the lamp output with a larger particle size phosphor is superior, introducing a conflict between cost and quality.

It was found that the greatest detriment to lumen output and maintenance was the very fine phosphor particles that were produced while reducing the phosphor to a particle size suitable for coating[2]. While the phosphors were being acid washed and in an aqueous slurry, it was convenient to also mill the phosphor to the particle size required for the final coating, at the same time removing many of the fine particles in the sub micron range. In this way, the milling was more accurately controlled so as to reduce the large particles, but at the same time minimize the production of fine particles—that is, to produce a phosphor with a rather narrow particle size distribution. By drying the phosphor in such a manner so as to prevent agglomeration, such as by spray drying[3], the phosphor no longer required milling in the lamp plant. The phosphor powder was then simply added to the lacquer and stirred to produce a paint.

The phosphor was milled in ball mills in the early period of wet milling, a slow and labor-intensive technique. At Westinghouse, the ball mills were replaced with a continuous-type vibratory mill. The phosphor was milled before acid washing.

When the particle size distribution becomes quite narrow, the adhesive properties of the phosphor deteriorates and the incidence of chipped coatings increases. To counteract this, a small amount of very fine aluminum oxide, 1 or 2%, is added to the paint. While effective, the aluminum oxide presents some difficulties in paint making, and in some cases calcium and/or strontium borates are added to the paint to aid phosphor adhesion.

Protective surface treatments for calcium halophosphates have been and still are being investigated, but as of this writing they have been unsuccessful.

Quality of Light

The quality of electrically produced visible light has always been a problem. Those of us who are old enough to remember when the only lighting was from incandescent lamps will recall that it was then common practice to take merchandise outside the store, into daylight, to judge its color. A criticism in the early days of incandescent lighting was that it was much too yellow compared to daylight. The fact is that there is much red and yellow but very little blue light in the incandescent lamp. This, of course, gives anything illuminated by this source a warm yellow glow.

Since the incandescent lamp was the only electric light source available, the public adapted to it, and in fact eventually it became the standard of comparison. Then the fluorescent lamp containing calcium halophosphate phosphors came onto the scene. Its light was rich in blue and green emissions but somewhat deficient in red, just the opposite of the incandescent lamp. To compound the felony, many of the early lamps were of so-called "daylight color"—after all, it was considered a good marketing tool to have a daylight color. But at a 6500K color temperature, this was a decidedly blue appearing lamp, especially when viewed next to incandescent light.

Eventually, the public's preference settled on the standard cool white, and in some cases the warmer "white" and "warm white" lamps. The warm white lamp color appeared similar to that of an incandescent lamp, but the color composition was different. As any artist knows, a given color can be produced in many different ways. So it is with light source color. The incandescent lamp chromaticity is the result of a large red component and decreasing amounts of orange, yellow, green and very little blue. The same appearing calcium halophosphate fluorescent lamp is composed of very little red, much orange and green, and some blue. The lamp colors appear the same, but colored objects viewed under the two lights can appear quite different.

Improved Color Rendering

The fluorescent lamp industry was cognizant of this shortcoming and attempted to correct it. This was originally done by replacing the calcium halophosphate phosphor with a blend of usually three or more phosphors, or blending other phosphors with calcium halophosphate phosphor. Together, in the proper proportions, they produced an identical color appearance that rendered objects viewed under its light similar to that viewed in daylight. The light emission from these lamps was distributed more evenly across the spectrum. This worked quite well as far as color rendering was concerned, but lamp efficacy in terms of LPW was reduced significantly because more of the light emission was now in the red and blue portions of the spectrum that have a lower luminous efficiency. Additionally, the phosphors' emission bands were rather wide and a portion of the emission extended into the infrared and ultraviolet regions, where the energy was wasted. Because of the lower light efficiency, more lamps and fixtures were required to obtain a given light level, and more power was required to operate an installation. Finally, because of the low volume and more costly phosphors, the lamp cost was significantly increased.

Tri-Band Lamps

By judicially selecting a blue, green and red phosphor with narrow band emission peaks corresponding to the peak sensitivity of the human visual system, a fluorescent lamp was developed that is very high in color rendering properties together with a lumen efficiency equal to, or even somewhat greater than, a standard calcium halophosphate fluorescent lamp. This was accomplished by using a new family of rare earth phosphors, which are ideal for this application. The term "tri-band" or "triphosphor" has been given to this type of lamp. Very little blue narrow band phosphor is actually required in the phosphor blend because most of it is supplied by blue mercury emission. The spectral power distribution of a cool white tri-band lamp is shown in Figure 5-2.

These lamps are more expensive because the rare earth elements are costly, but they do provide excellent color rendition. Color contrast

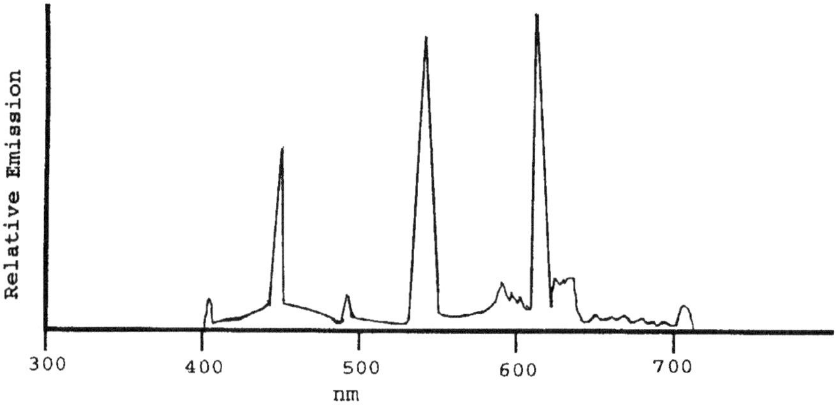

Figure 5-2. Cool white tri-band or triphosphor lamp emission.

of scenes viewed under these lamps is enhanced, causing the viewer to perceive the scene as brighter and cleaner than when viewed under a standard calcium halophosphate lamp. An added plus of the rare earth phosphors is that they are more stable under the lamp operating conditions and deteriorate less with use (better lumen maintenance), so that there is less change in light level during the life of the lamp.

Double Coating

A rather ingenious technique is used to reduce the rare earth phosphor cost in standard T12 tri-band lamps. The inside of a typical fluorescent lamp is coated with a uniform layer of phosphor at an optimum thickness. A heavier or lighter coating will result in reduced efficiency. However, most of the light is generated in the portion of phosphor facing the inside of the lamp, where the ultraviolet emission is generated. By first coating the lamp with a thin layer of low-cost calcium halophosphate phosphor, and then over-coating with a rare earth phosphor blend of the same color, a lamp results with properties essentially the same as the rare earth lamp, but with only half the amount of high cost rare earth phosphors. A cross section of a double-coated tri-band lamp is shown in Figure 5-3.

Small-diameter Lamps

Standard fluorescent lamps are predominantly T12 lamps, which stands for $12 \div 8 = 1.5$ inches in diameter. A new generation of T8 and

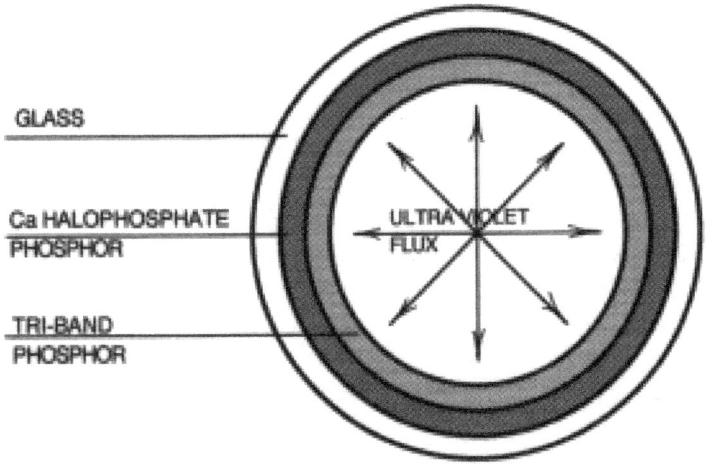

Figure 5-3. Cross-section of a double-coated tri-band fluorescent lamp.

T5 rare earth phosphor lamps (one-inch and 5/8-inch diameter) are now being manufactured. These produce light more efficiently than the present T12 lamps. The smaller diameter results in an increased ultraviolet flux impinging upon the phosphor surface, which is detrimental to calcium halophosphates. This problem is overcome by using the higher-cost rare earth phosphors, which are more tolerant of the high UV flux. The smaller diameter of the T8 and T5 lamps reduces the quantity of phosphor required to coat the lamp, thereby offsetting some of the higher cost. Increasing use of rare earth phosphors will eventually be reflected in decreased demand for calcium halophosphate phosphors.

Rare Earth Phosphors

High-purity rare earth elements became available in commercial quantities in the late 1970s. The development of rare earth activated phosphors followed. The rare earth activated phosphors are noted for their well-defined narrow band-width emission. The activators useful for the tri-band high color rendering index (CRI) lamps are europium in the 2+ valence state (Eu^{2+}) for the blue, terbium in the 3+ valance state Tb^{3+} for the green and Eu^{3+} for the red. Some of the rare earth phosphors with Eu and Tb activators are described below. There are many other phosphors with these and other rare earth activators.

The development came gradually. The red europium-activated yttrium oxide was first, followed by blue Eu^{2+} activated strontium chlorophosphate. The first commercial tri-band lamps were a blend of the above blue and red with manganese-activated zinc silicate as the green. This green had a wide band-width, but nevertheless produced a high-CRI lamp. Lumen output was equal to that of a standard calcium halophosphate lamp. The green deteriorated more rapidly than the rare earth phosphors, however, and the lamp slowly changed color during life. This was particularly marked near the electrodes and the lamps frequently exhibited pink ends later in life. Various techniques were tried to counteract this, but with variable success. The addition of a small amount of antimony trioxide to the coating improved the maintenance of the green and was beneficial. The problem was eliminated with the development of the green rare earth phosphor cerium terbium magnesium aluminate, which in turn has been mostly replaced by lanthanum cerium terbium phosphate.

Red Europium-Activated Yttrium Oxide (Y_2O_3:Eu)

This is a simple oxide phosphor and can be prepared by simply mixing the two oxides and firing in air at about 1350°C for 12 hours. The brightness of phosphor produced by this simple technique is generally about 5% below that attainable with some minor modifications. Three workable procedures are shown below:

1. Dry mix the rare earth oxides together with a small amount of fluxing agent, such as calcium, that is retained but is not detrimental to quality. Fire at 1350°C.

2. Mix the yttrium and europium as a solution and coprecipitate. A small addition of calcium is helpful. Typically the oxides are dissolved in HCl or HNO_3 and precipitated as the oxalate. The oxalate decomposes to the oxide when heated. Fire at 1350°C.

3. Mix the oxides together with zinc chloride, which acts as a flux, and then fire until all the zinc is vaporized. A second firing is usually required to remove all of the zinc chloride. This is not a preferred technique because of the copious zinc chloride emission during firing.

All three methods will produce a good phosphor, but the first two are most commonly used, depending upon the preference of the manufacturer.

This phosphor is extremely stable, depreciating very slowly during lamp operation, and it is very resistant to damage during milling. It is used as the red component of the tri-band lamp and color cathode ray tubes. For cathode ray use the Eu content is reduced. The SPD curve of this phosphor is shown in Figure 5-4.

Green Magnesium Cerium Terbium Aluminate ($Mg(Ce,Tb)Al_{11}O_{19}$:Ce,Tb)

This is an efficient and stable phosphor, but it is somewhat difficult to prepare. The aluminum oxide must be of high purity, manufactured so as to be reactive, and conform to a rather exacting particle size specification. The components do not mix well in the dry state and typically are mixed as well as possible, fired to obtain a preliminary reaction, milled to complete the mixing and again fired. Firing is in a slightly reducing atmosphere and at a high temperature, around 1500°C, for a long period of time. This phosphor replaced the zinc silicate used in early versions of the tri-band lamp. The SPD curve of the terbium activated phosphor is shown in Figure 5-5.

Green Lanthanum Cerium Terbium Phosphate ($(La,Ce,Tb)PO_4$:Ce,Tb)

This phosphor has mostly replaced the green aluminate because it is prepared at a firing temperature of about 1200°C, and a firing time of about three hours. This is significantly less costly than the aluminate and the quality is comparable. Firing is in a slightly reducing atmosphere. The SPD is similar to that of the green aluminate.

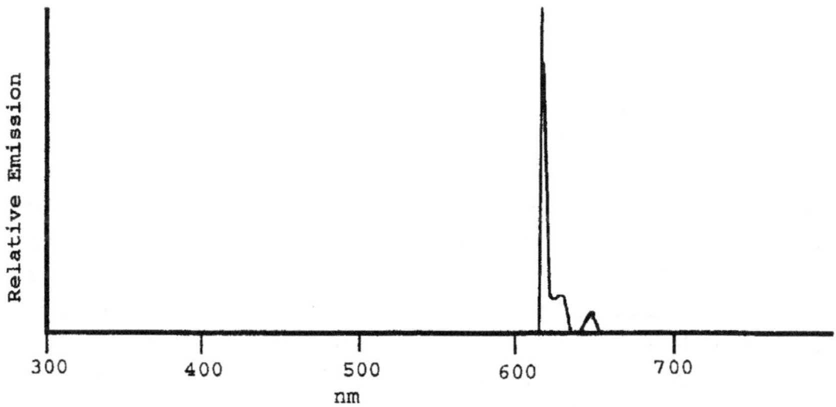

Figure 5-4. Spectral power distribution curve for a red europium-activated yttrium oxide phosphor.

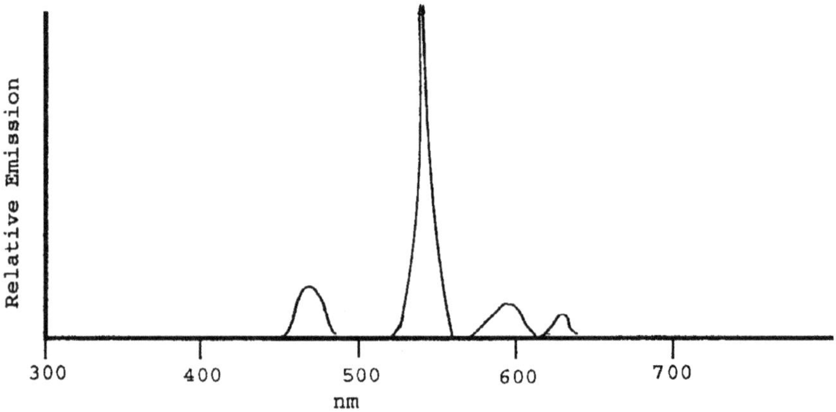

Figure 5-5. Spectral power distribution curve for a green magnesium cerium terbium aluminate phosphor.

Blue Europium-Activated Strontium Chlorophosphate ($Sr_{10}(PO_4)_6Cl_2:Eu$)

This is probably the first useful narrow band blue emitting phosphor. The emission is a rather narrow band at 450 nm and it can be used in the tri-band lamp. The preparation is somewhat unique in that an excess of strontium chloride, which fuses during firing, is added to the raw mix as a flux. The fired cake is a hard mass of phosphor particles contained in the fused chloride. The excess strontium chloride is dissolved to release the phosphor powder. A slightly reducing atmosphere is required during firing.

Blue Europium-Activated Barium Magnesium Aluminate ($BaMg_2Al_{16}O_{27}:Eu$)

This is another efficient and stable phosphor that is now generally used as the blue component of the tri-band lamp. Preparation is similar to the green aluminate phosphor in that dry mixing is not complete and two firings are required with milling in-between the two firings. When prepared by simply mixing the components, a firing temperature of around 1450°C is required and a rather large particle size phosphor is produced. By adding a small amount of a suitable flux, such as aluminum fluoride, the firing temperature is reduced to around 1250-1300°C and the product has a more desirable small particle size. A slightly reducing firing atmosphere is required. The SPD curve, shown in Figure 5-6, is similar to that of the europium-activated strontium chlorophosphate phosphor.

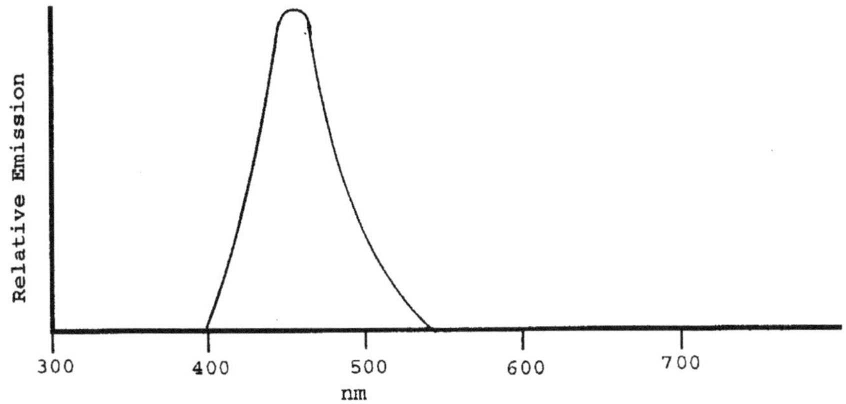

Figure 5-6. Spectral power distribution curve for a blue europium-activated barium magnesium aluminate phosphor.

High Pressure Mercury Vapor Lamp Phosphors

The high pressure mercury vapor lamp (HPMV) was developed for practical use following World War II, more or less paralleling the development of the fluorescent lamp. It was not as efficient as the fluorescent lamp, but was significantly more efficient than the incandescent lamp. A major advantage was the availability of high-wattage lamps capable of illuminating large areas such as highways, streets and parking lots, with a small number of high-mounted luminaires. Fluorescent lamps with their low per unit light availability, albeit more efficient, required a significantly greater capital installation cost. The major disadvantage of the HPMV lamp was the very poor light quality. Most of the emission is in the green and blue region resulting in extreme color distortion.

Lamp Emission

Mercury lamps are constructed with a small quartz tube with electrodes, containing mercury under high pressure when operating hot. The arc tube is contained in a special heat resistant glass outer that is quite hot when operating. Detailed information about HID lamps is available elsewhere in this book.

Mercury lamps generate a large amount of wasted ultraviolet radiation in addition to the desired visible radiation. It was desirable to improve the color characteristics of the lamp. One obvious method was to make constructive use of the wasted ultraviolet. By coating the inner surface of the outer bulb with a phosphor, the wasted ultraviolet could be utilized to produce light in regions of the spectrum that would improve the color rendering properties of the lamp. A phosphor coating on the inside of the outer envelope would not be subjected to the mercury of the arc and would, in this context, experience a less rigorous environment. But the outer envelope of this lamp is quite hot, about 360°C, and the emission characteristics of most phosphors are temperature dependent. There were only a limited number of phosphors that were efficient at the lamp's operating temperature.

Magnesium Fluorogermanate:Mn

The first practical phosphor developed for this application was a deep-red-emitting manganese-activated magnesium fluorogermanate (MFG), $Mg_4FGeO_6:Mn^4$. This phosphor emitted multiple bands of deep red and was efficient at the elevated temperature encountered. The CRI was improved, but because of the deep red phosphor emission, and phosphor absorption of some blue lamp emission, the LPW actually decreased a little. MFG was an expensive phosphor because germanium was costly and the manufacturing process, described below, was lengthy:

1. Weigh the components into a tumbling type mixer and mix.
2. Hammer mill.
3. Again tumble mix and hammer mill. Mixing is still incomplete because the components pack and do not flow freely.
4. Fire in air for two hours at 1250°C. The partial reaction eliminates packing during mixing.
5. Hammer mill and tumble mix to complete the mixing.
6. Fire in air for 12 hours at 1250°C.
7. Ball mill.
8. Refire in air for 12 hours at 1250°C.

Strontium Magnesium Phosphate:Sn

About 1970, MFG was replaced by the much cheaper orange-emitting tin-activated strontium magnesium phosphate (MP) phosphor

$(Sr,Mg)_3(PO_4)_2$:Sn)[5]. This phosphor improved both CRI and lamp output because the emission did not extend significantly into the deep red region, and did not absorb visible radiation.

MP manufacture was rather conventional except that the firing atmosphere had to be slightly reducing, but very controlled. If the atmosphere become non-reducing, the tin activator oxidized to the +4 valance state, resulting in a very white cake with no fluorescence. But if the atmosphere become excessively reducing, some tin was reduced and a dark, poor fluorescing product was produced.

Yttrium Vanadate:Eu

The MP phosphor was replaced by europium activated yttrium vanadate (YVO_4:Eu), a rare earth phosphor with a narrow band emission in the orange-red region, peaking at 610 nm.

This phosphor was prepared by a coprecipitation process. Vanadium pentoxide and the rare earth oxides were dissolved in hydrochloric acid and precipitated with sodium hydroxide containing some borate, the borate acting as a flux during firing. The precipitate was washed to remove the sodium and chloride, then filtered and dried. The dried cake was milled and fired in air at 1200°C, washed to remove any residual sodium chloride, and then dried.

Yttrium Vanadate Phosphate:Eu

The europium activated yttrium vanadate phosphate $Y(V,P)O_4$:Eu phosphor was favored by Westinghouse over the YVO_4:Eu until the Westinghouse Lamp Division was purchased by Philips Lighting. The YVO_4:Eu was brighter initially, but the $Y(V,P)O_4$:Eu had better maintenance. A disadvantage of the $Y(V,P)O_4$:Eu was that it is insoluble in mineral acids, making it impractical to reclaim the rare earth values if a batch did not meet quality standards.

The manufacturing process for $Y(V,P)O_4$:Eu was more complicated. A coprecipitated YPO_4:Eu was made by dissolving the rare earths in hydrochloric acid and precipitating with phosphoric acid. After washing, filtering, and drying, the YPO_4:Eu precipitate was mixed equally with YVO_4:Eu precipitate and fired in air at 1200°C.

As of this writing, the HPMV lamp, while still available, is becoming obsolete. The metal halide and high pressure sodium lamps are both much more efficient and offer better color rendering.

Ultraviolet Phosphors

Skin tanning is a major application for ultraviolet-emitting fluorescent lamps. Other uses are for medical, photocopying, fluorescent displays and photochemical reactions. The last three uses are minor and will not be discussed here. Photocopy lamps may be green, blue or ultraviolet, depending upon the application. There are a large number of ultraviolet-emitting phosphors, most finding little or no application. Only several of the better known will be briefly discussed.

	Ultraviolet Phosphors		
Crystal	*Activator*	*Emission Peak (nm)*	*Width at 50% (nm)*
Calcium Orthophosphate	Tl	310	55
Barium Disilicate	Pb	355	40
Strontium Fluoroborate	Eu	370	16
Strontium Hexaborate	Pb	290	22

Thallium-activated Calcium Orthophosphate $Ca_3(PO_4)_2$:Tl

This phosphor was manufactured by Westinghouse from about 1950 to 1970. It was used in the first commercially manufactured erythemal lamp, and was used medically for the treatment of psoriasis.

The manufacturing process was routine, consisting of mixing the components and firing in air. Care was necessary to contain a small amount of poisonous thallium emission during firing.

Lead-activated Barium Disilicate $BaSi_2O_5$:Pb (BDS)[6]

BDS phosphor was first manufactured commercially in the late 1940s as a long-wave ultraviolet phosphor for photocopying. In the 1970s, as the mechanism of skin tanning became known, it found major application in tanning lamps. The manufacturing process was routine. Firing was in air.

The maintenance of this phosphor was relatively poor and resulted in short useful lamp life. In the late 1980s, a process for applying a protective coating of aluminum oxide to the phosphor crystals was developed by Nichia. This improved the maintenance so that it is now competitive with SFB (see below).

Europium-activated Strontium Fluoroborate $SrFB_4O_{6.5}$ (SFB)

SFB phosphor has excellent maintenance and was introduced at about 1970 as an improved black light phosphor. It found use in tanning lamps as a replacement for BDS, but found resistance because of the high cost.

SFB must be manufactured with care, or the borate will fuse into a glass during firing, rendering it useless. Osram Sylvania manufactured SFB by a combination of coprecipitating and mixing of components[7]. Westinghouse used a process of dry-mixing the components and multiple firing. In both cases, processing is very exacting. The final firing must be in a reducing atmosphere for the Eu to be in the +2 valence state. The finished phosphor must be washed to remove unreacted borate. Failure to do so results in residual free borate fusing with the glass surface during the lamp lehring operation that removes the paint binder. Upon cooling, the difference in the coefficient of expansion results in catastrophic failure of the glass.

Lead-activated Strontium Hexaborate

Tanning lamps are composed of mostly UV-A emission but require a small amount of UV-B emission. The U.S. Federal Drug Administration defines UV-A as radiation between 321 and 400 nm, and UV-B as radiation between 260 and 320 nm. The SFB phosphor is a narrow-band emitter that peaks at 370 nm, and BDS is a broad-band emitter with a peak at 355 nm.

The lead-activated strontium hexaborate is often used as the UV-B component in tanning lamps made with the narrow band SFB phosphor. Lead-activated barium zinc magnesium silicate can also be used. The broad band emission of BDS extends slightly into the UV-B region and does not require additional UV-B phosphor. The BDS is now the predominant tanning lamp phosphor because of its cost advantage.

Specialized Applications

There are a number of specialized applications for fluorescent lamps that are not commercially important to the lamp industry but are, nevertheless, quite interesting. Only a sampling will be touched on here.

Infrared Lamp

While of no real commercial importance, these lamps were manufactured by Westinghouse for a short period of time. The phosphor was iron-activated lithium metaluminate, $LiAlO_2:Fe^{3+}$ [8]. It emitted several peaks in the near infrared range of about 675-875 nm, with a tail into the deep red. This made the lamp color appear a deep dull red. The phosphor was prepared by the conventional technique of mixing the components and firing in air. Two firings with milling in-between were necessary. The finished phosphor required washing in a weak acid to remove any unreacted lithium. Failure to do so resulted in lithium reacting with the glass during lehring followed by disintegration of the glass upon cooling.

Philips Lighting discontinued the phosphor after purchasing the Westinghouse Lamp Division. Osram Sylvania still lists this phosphor.

The Bilirubin Lamp

The Bilirubin lamp is coated with the deep blue-emitting phosphor, $Sr_5Cl(PO_4)_3:Eu$, that emits in a very narrow band at about 450 nm. It is used to irradiate new born infants, usually premature, who are jaundiced. The blue light penetrates the skin and aids in the decomposition of excess bilirubin in the blood stream.

Sorting Produce

Another interesting application is in the sorting of agricultural produce. Lamps with various emissions are used depending upon the product to be sorted and the type of separation desired. The reflection of the light from the product is monitored, and separation is made based upon differences. As an example, dark particles are removed from rice using the reflection of a deep blue lamp. An infrared emitting lamp is used to sort coffee beans using a difference in reflectivity that cannot be seen with the human eye.

Electroluminescent Lamps

An electroluminescent phosphor is a material that fluoresces when subjected to a changing electric field. In the early 1960s, there was a rather extensive effort to capitalize on this phenomenon. The long-range goal was to produce an area light source for applications such as lighted ceilings and walls. Experimental panels were produced in 1- and 2-square-foot sizes, as well as gimmicks such as lighted window shades. Excellent read-out panels were also made on an experimental basis, but that was in the age of analog read-out devices, when digital read-outs were of little interest.

Electroluminescent Lamp Construction

The lamps were essentially flat electrical condensers, with one plate made of an opaque reflecting conductor, and the other plate a transparent conductor. A transparent insulating material containing an electroluminescent phosphor was placed between the plates. An alternating voltage applied across the plates activated the phosphor, emitting light as the voltage fluctuated with each cycle. Light increased with increasing voltage and frequency. Most testing was done at 600V and 3000 Hz. If the voltage was excessive, arcing occurred, causing dead spots and in extreme cases, complete destruction of the lamp. The flexible lamps were basically a phosphor-impregnated plastic sheet with a vaporized aluminum coating as one plate and a transparent conducting coating as the other plate. The "glass" lamp was a glass plate with a conducting tin oxide coat covered with a phosphor imbedded plastic coating. A vaporized aluminum coat was applied as the reflecting conductor. The so-called "metal" lamp was a metal plate covered by a thin insulating layer of low melting point glass, a layer of phosphor imbedded in glass, another thin glass insulating layer, and a vaporized, conducting tin oxide coat. All of these lamps were susceptible to moisture damage, so they all required a final protective coating. A metal lamp night light was produced and sold as a consumer product. The night lights were green, but experimental white lights were made by blending yellow, blue and green phosphors. The metal lamp was also used in the instrument panels of military aircraft. An electroluminescent night light is again being advertised in consumer catalogs.

The Phosphors

The electroluminescent phosphors were mostly zinc sulfides activated by copper and a halide. The color was green when activated by copper and chlorine. Blue was produced by replacing the chlorine with bromine or iodine. Many variations from deep green to bright blue were obtained with various combinations. Yellow was obtained by using manganese as a coactivator. Red was obtainable by using zinc selenide, but was not developed because of the toxicity of selenium. The electroluminescent properties were developed by firing the phosphor with a large excess of copper sulfide so that the zinc sulfide crystals contained tiny crystallites of copper sulfide. In the lamp, electrons in the copper sulfide move back and forth as the electric field changes, producing fields inside the crystal, exciting the activators to produce luminescence. After firing, a large excess of copper sulfide on the surface of the zinc sulfide crystals was removed by washing in sodium cyanide solution.

The phosphors were fired in a neutral or reducing atmosphere to prevent oxidizing the sulfur and producing useless oxides. The halides were important in determining the color and had to be introduced quite accurately. One method was to add the halides as ammonium salt and fire in a covered tray with a protective atmosphere. A small amount of excess sulfur was typically added. But there was a variation in cover integrity from tray to tray, resulting in color variation. Tight color control was obtained by firing in open trays and introducing the halide in the gaseous form as the hydrogen compound. Nitrogen as used as the carrier gas. Typically, some hydrogen sulfide gas was also included. Electroluminescent phosphors are still listed by Osram Sylvania.

The Future of Lamp Phosphors

Fluorescent Lamp Phosphors

The fluorescent lamp is a well-established light source that will find application long into the future. New, more efficient light sources may be discovered, but the market for fluorescent lamps will continue just as the incandescent lamp continues to exist in spite of more efficient competitors. As long as fluorescent lamps are made, there will be a market for lamp phosphors.

The demand for various phosphors changes, but old phosphors never seem to die, with the exception of the toxic zinc beryllium silicate. Many become almost insignificant, but there always seems to be some minor application that continues indefinitely. Phosphor manufacturers continue to supply these phosphors, although often on a custom basis.

The calcium halophosphate phosphors have been the main stay of the lamp industry for over 50 years. The rare earth phosphors are now making inroads and will continue to grow in importance, but the low cost and versatility of the calcium halophosphate phosphors gives them a rather compelling advantage, guaranteeing they will be of major importance for many more years.

The major disadvantage of the calcium halophosphate phosphors is the poor lumen maintenance under high flux ultra violet radiation, further aggravated by the lamp industries' moving in the direction of smaller diameter lamps. Much effort is being devoted to develop protective coatings to overcome this problem, which ultimately may lead to a solution. Since the relatively poor color rendering characteristic of the calcium halophosphates is not detrimental for many applications, its future use is well assured. The demand for rare earth phosphors will continue to grow, and will share a major status with the calcium halophosphates.

High Pressure Mercury Vapor(HPMV) Lamp Phosphors

The demand for phosphors for this line of lamps will gradually diminish. New installations where HPMV lamps were used in the past, now incorporate the superior metal halide and high pressure sodium vapor lamps, neither of which need phosphor coatings for color rendering improvements. Phosphors for HPMV lamps will eventually become specialty items.

New Phosphors

It is highly unlikely that a new, commercially important lamp phosphor system will be found. More of a possibility is that a known, but unused, phosphor may be developed for commercial application. There has been extensive exploration of known crystal systems, incorporating most elements of the periodic table, but to the point of diminishing returns.

The more promising direction for future development is in the adoption of coatings that will protect phosphor crystals from the harsh operating environment of the fluorescent lamp. This could be particularly beneficial for calcium halophosphate phosphors.

References

1 McKeag and Ranby, U.S. Patent 2,488,733.
2 Van Broekhoven and Repsher, U.S. Patent 2,255,373.
3 Vodoklys, U.S. Patent 3,023,339.
4 Thorington, U.S. Patent 2,748,303.
5 Koelmams et al., U.S. Patent 3,110,680.
6 Ginther, U.S. Patent 2,499,307.
7 Chenot, U.S. Patent 3,431.215.
8 Hummel and Sarver, J. Electrochemical Society 111 252, (1964).

Chapter 6

The Ballast

by Raymond Kane

Functions and Features

Gaseous discharge arc lamps will produce light when the gas they contain is made to conduct by applying a sufficiently high voltage across the electrodes. The gas becomes ionized and an electric arc is produced. The arc has a negative impedance, which means that unless some means is provided to control its current flow, the equivalent of a short circuit occurs, and the current will continue to increase until the lamp destroys itself. To prevent this from occurring, a current limiting device, or "ballast," must be incorporated into the circuit.

The ballast must be designed to perform several critical functions:

1. It must limit the current to the range specified by the lamp manufacturer because the lamp is essentially a constant voltage device. The desired result is to have the lamp operate at its rated wattage so that it will provide the proper amount of light.

2. The ballast must provide sufficient open circuit voltage to initiate and sustain the arc. In many instances, this voltage is not available at the source, so the ballast must provide some means to transform the input voltage to a higher value to make certain that the lamp will start.

3. On every half-cycle, as the current wave passes through zero, the electron density diminishes and the arc extinguishes. The ballast must then allow ionization to re-occur very rapidly in order to re-

start the lamp. This is done by controlling the amplitude and the phase angle of the open circuit voltage wave.

It is also desirable to regulate the lamp wattage. Variations in the voltage supplied to the lamp can alter the lamp's wattage and consequently, its light output. The ballast will absorb some of this variation, which helps to maintain a more constant output. Because the lamp and the basic ballast constitute an inductive circuit, they require more current to operate than does an incandescent lamp of the same wattage. To compensate for this phase shift effect, a capacitor is often built into, or supplied with the ballast, so as to increase the system's overall power factor.

The ballast, as part of the lamp's operating circuit, consumes power, as does the lamp itself. This ballast power produces no light, so it must be viewed as a loss. Therefore, to establish the efficiency of the overall lighting circuit, the ballast losses must be determined. If 85% of the total power is used by the lamp and 15% is used to heat the ballast, then the efficiency of the lighting system is considered to be 85%. For example, a 40W fluorescent lamp may operate with a ballast that consumes 5 watts of power. The total energy delivered to the lighting system is 45 watts. Also, it must be understood that all of the 40 watts delivered to the lamp is not converted to visible light.

Construction of Electromagnetic Ballasts

The simple reactor-type, electromagnetic lamp ballast consists of a core and coil assembly, to which is often added a capacitor to correct the line power factor. Other electromagnetic lamp ballasts are basically a modification of the reactor ballast's construction. Figure 6-1 shows the construction of a fluorescent lamp autotransformer lamp ballast.

Every ballast design must be a compromise of size, weight, shape, performance and cost. It must have a long life, produce a minimum of hum, and have a low power loss. There is a constant demand by luminaire manufacturers for ballasts of increasingly smaller dimensions as they seek to make the luminaire units smaller.

The Design of the Core and Coil Assembly

The ballast core is composed of many layers of electrical steel laminations. The laminations usually range from 0.010 to 0.025 inches in

Figure 6-1. Typical discharge lamp ballast construction. Courtesy of Advance Transformer Company.

thickness and are insulated from each other by a very thin oxide layer on their surface. Very often, the oxide scale formed during rolling and annealing provides this layer. The lamination sheet is rolled in a special fashion, which orients the grains in a pattern that enhances its flux-carrying capacity, and reduces the core losses. Hot-rolled steel was originally used to make the sheet, because it lent itself to controlling the grain direction. However, a transition to less expensive cold-rolled steel was made after orientation improvements were designed into the final core assembly to reduce these losses. Later, a high temperature stress relief was incorporated into the cold-rolled process to achieve a further loss reduction.[1]

The ballast coil is made of copper or aluminum magnet wire, which is wound around a mandrel carefully and uniformly, to provide a uniform flux path. The wire is coated with a varnish of polyurethane or polyester to prevent any short-circuiting between the coil turns.

Core and Coil Losses

There are two major power losses in the core. The first comes from the eddy currents, which are minimized by building the core with laminations, instead of using a single block of steel. The other is the hysteresis loss, which is minimized by removing the carbon impurities from the steel, and by adding additives such as silicon to it. There is another significant ballast loss that is produced by the I^2R loss in the wire. All of these losses tend to increase, as the demand to reduce ballast size continues.

Shell and Core Designs

There are two fundamental electromagnetic ballast constructions. The shell-type (see Figure 6-2), wherein the coil is contained within the housing of the core by virtue of being wrapped around the core's inner leg, and the core-type construction (see Figure 6-3), in which the bulk of the coil assembly is exposed because it is wrapped around the core's outer leg.[2] Because of its greater containment, the shell-type construction is the one that is most predominantly used, particularly for fluorescent lamps.

To control the flux density and the voltage waveform, the core structure can be modified to include an air gap, magnetic shunts, slots and splines, all of which will alter the cross-sectional area of the magnetic flux path.

The Potting Process

In addition to the insulation provided for the coil wire and the core laminations, another step has been added, whereby the entire ballast assembly is completely impregnated by dipping it into a liquid "potting" bath for a few minutes, then allowing it to dry in a conveyor-oven. Potted ballasts are usually quieter and run cooler than unpotted ones.

The potting process has been improved by placing the ballasts into a vacuum chamber from which all moisture and air is removed, then pouring the potting compound into the chamber. Within a few minutes, all of the spaces and crevices within the ballast assembly become completely filled. The excess liquid is then drained off and the assembly is allowed to dry, usually in a batch-type oven.

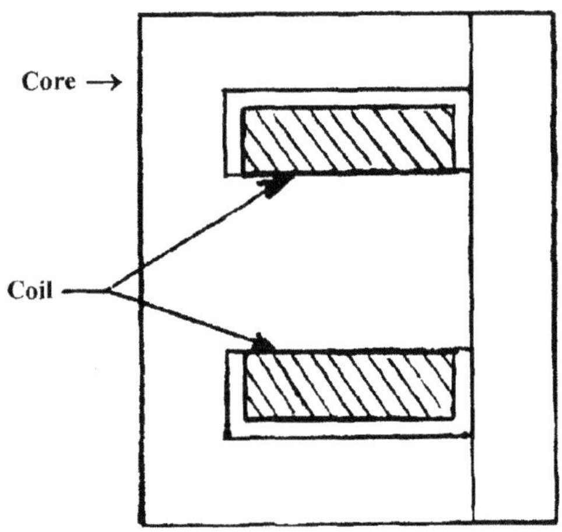

Figure 6-2.
A shell-type ballast construction.

Although it is a good insulator, the potting compound must also be a good conductor of heat and must have a fairly low coefficient of expansion. New developments in silica resins and fillers have made them

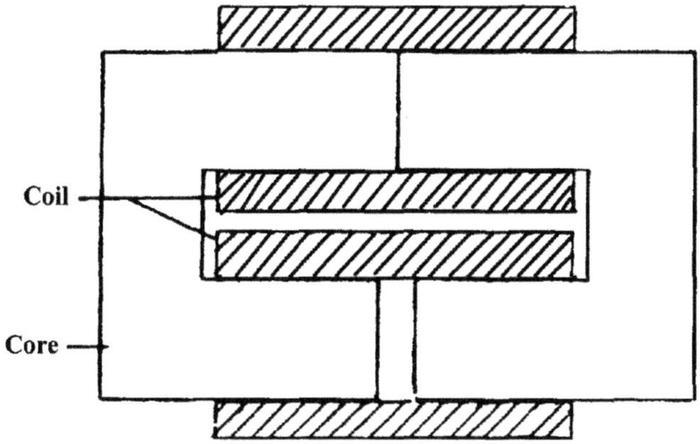

Figure 6-3. A core-type ballast construction.

more suitable than the older petroleum-based compounds, which tended to expand and eventually leak out of the ballast casing. The newer resins harden during ballast processing and do not soften or become molten in service. The potting compound helps to transfer the heat generated in the coil assembly to the core where it is dissipated.

General Characteristics

There are many characteristics that influence the design and performance of ballasts used with discharge lamps. These traits will vary with the types of lamps with which the ballasts are used; however, whether the lamps are fluorescent, mercury vapor, metal halide or high pressure sodium, it is helpful to become familiar with the following attributes:

Crest Factor

The wave shape of the alternating current passing through a discharge lamp is not sinusoidal. The degree of distortion depends upon the type of ballast used. The accepted criterion for the measurement of this distortion is called the crest factor, which is the ratio of the peak value to root-mean-square value of the lamp current. The crest factor has a minimum value at 1.41 for a pure sine wave.

Although the input voltage supplied to a ballast is generally a sine wave, the waveshapes of the secondary voltage and current are distorted because of the capacitive and inductive components in the output circuit, which is why the crest factor becomes greater than 1.41.

A typical distorted current waveform which results in increasing the crest factor is shown in Figure 6-4.

The crest factor increases as the current wave becomes more peaked. Its measured value does not identify the exact wave shape, but it is a useful measurement for evaluating the performance of discharge lamps.

Tests by Martt & Smith show that crest factor values can be changed by varying the line voltage and/or the lamp voltage.[3]

Figure 6-5 shows such variations in crest factor for a 400W mercury vapor lamp.

Many of the high intensity discharge mercury lamp ballasts in use today are of the constant wattage(CWA) design, and these have a crest fact of about 1.8. Metal halide and high pressure sodium ballasts, on the other hand, are usually about 1.65.

Effect of Crest Factor on Lumen Maintenance

Tests conducted by several lamp and fixture manufacturers showed that a greater rate of lumen depreciation occurs when the lamps are operated on ballasts having a high crest factor. Similar work conducted by the Wide Lite Corporation on 400W mercury vapor lamps demonstrated that increasing the crest factor will produce a significant decrease in mean lumens (see Figure 6-6).[4]

Studies by Smith & Matz established that lumen maintenance could be improved on lamps operating on ballasts of relatively high crest factor (up to 2.0) if the lamps were made with oxide-coated electrodes rather than the customary (for that period) thoriated tungsten electrodes.[5] They also found that increasing the lamp voltage crest factor improved the cold starting (-20°F) reliability of the lamps when the RMS open circuit voltage is reduced.

Effect of Crest Factor on Service Life

Noel & Martt reported that the crest factor can go as high as 2.7 during starting.[6] Such wave distortion requires a longer lamp stabilization period than normal. Also, the electrode will have to provide the maximum current of the waveform which will cause the electrode temperature to rise. More rapid evaporation of the electrode will occur, resulting in earlier arc tube blackening and a reduction in lamp life.

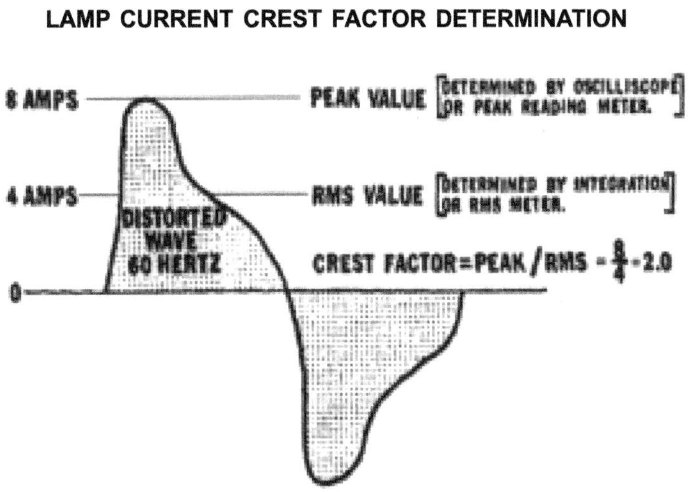

Figure 6-4. A typical distorted current waveform from which the crest factor is determined. Courtesy of Wide Lite Corporation.

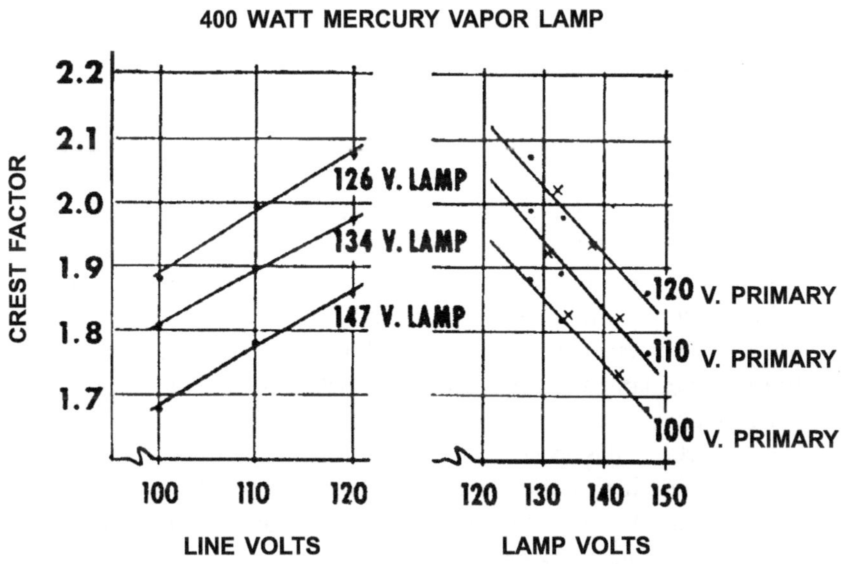

Figure 6-5. Crest factor values as a function of line and lamp voltages as measured on 400W mercury lamps (from Martt and Smith, 1958). Courtesy of the Illuminating Engineering Society.

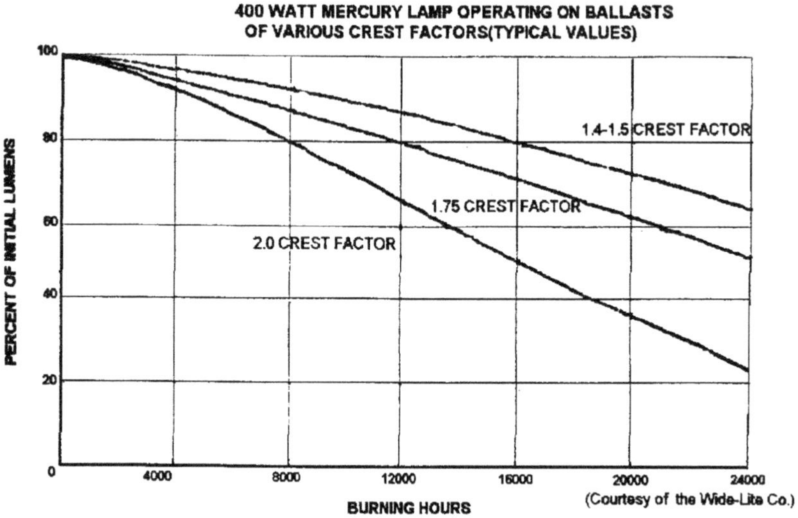

Figure 6-6. Crest factor effect on the lumen maintenance of a 400W mercury vapor lamp.

Regulation

Ballast regulation is defined as the resultant change in lamp wattage when the line voltage to the ballast is changed.[7] Furthermore, any change in lamp wattage is usually accompanied with a proportional change in lamp light output. With metal halide lamps, the wattage and light output variation are proportional only up to a maximum change of +/- 10%. When the line voltage drops below 90% of its rated value, the temperature of the discharge may be reduced enough to cause one or more of the additive compounds to de-ionize, thereby not only reducing the light output, but dramatically changing the color as well.

Many ballast circuits have been developed to operate high intensity discharge lamps, always consistent with the need for good regulation. However, improvements in regulation are most often achieved by increasing the ballast cost.

Wherever high intensity discharge lamps are to be used, consideration must also be given to the expected variation the input voltage will have on the lighting system.

Most new systems are regulated to operate between +/- 5% of nominal line voltage. Older systems may vary as much as +/-10% or more.

Power Factor

Power factor as defined for electrical wave shapes is the ratio of the actual power that is consumed in the circuit which is measured in watts, to the power apparently taken from the line which is measured in volt-amperes. A watt-meter is used to measure actual power, whereas a volt-meter and an ammeter are used to measure apparent power.[8]

In those circuits that are totally resistive, such as electric heaters and incandescent filament lamps, the actual power and apparent power are the same. There, the power factor has a value of 1.00.

In those circuits that are not totally resistive, such as inductive ballast circuits and circuits used to operate motors, having a low power factor (LPF) can be very detrimental and expensive. This is because the demand for volt-ampere power increases substantially, and an extra premium or penalty will be imposed by the electric utility company. Furthermore, the useful load that the circuit wiring and its accessories can handle is reduced significantly.

Lamp circuits which use a simple inductor or a leakage reactance transformer ballast have a low power factor ranging from 0.30- 0.50.

A lagging power factor can be corrected quite simply by connecting a capacitor of the proper size directly across the incoming AC supply. The capacitor draws a leading phase current that nearly compensates for the lagging phase current of the ballast. The total circuit power factor can be raised to as much as 0.95, which can reduce the current demand by as much as 40%.

Figure 6-7 shows three fluorescent lamp current vector diagrams with their resultant lamp power factor angle (θ); it is the angle subtended by the lamp current (I) and the open-circuit voltage (V_{oc}).

Ballasts designed to a power factor greater than 0.90 are designated as "high power factor" (HPF) ballasts. Those between 0.50 and 0.90 are called "normal" power factor (NPF).

In those cases, where the high starting current does not over-burden the installation's wiring capacity, and the number of ballasts involved is small, the lower cost of a normal power factor ballast may be justified.

Noise

The mere presence of magnetic materials in all ballast circuits will produce some degree of audible noise or "hum" which comes from the harmonics of the 60 Hz input power. Most of this vibration is caused by

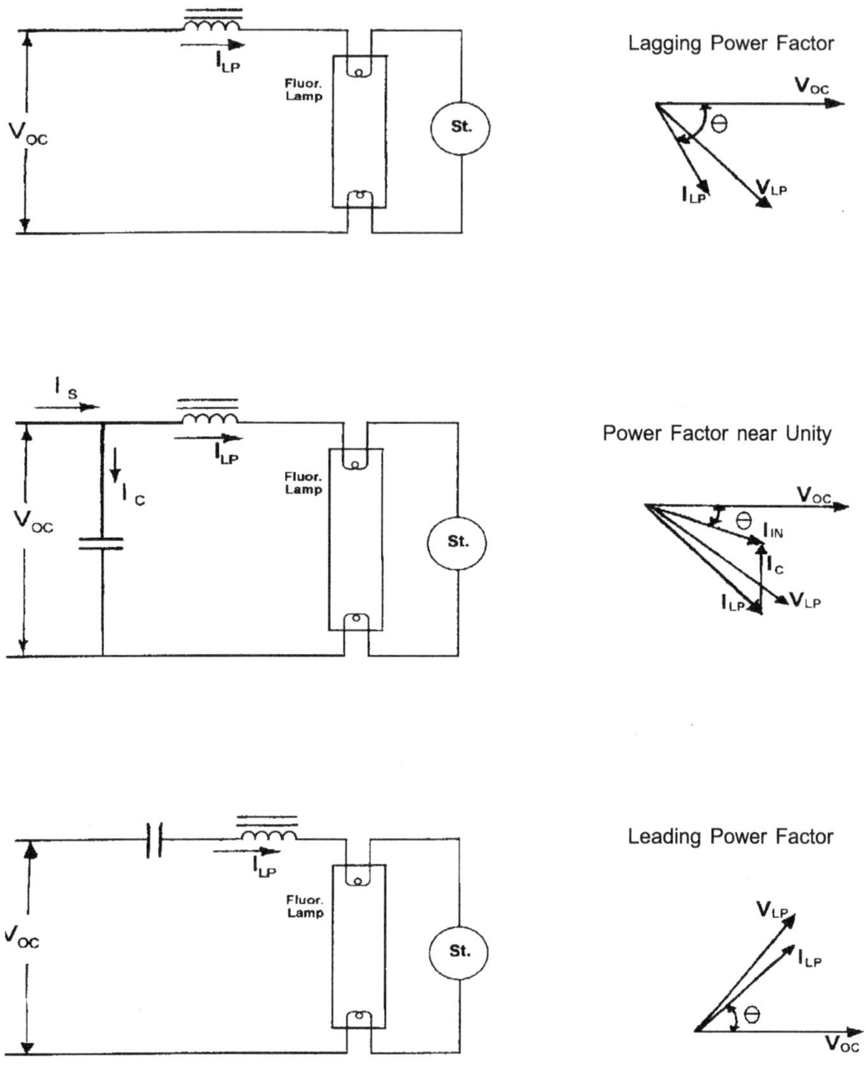

Figure 6-7. Lamp vector diagrams showing power factor angle (θ).

the expanding and collapsing magnetic field in the laminated core of the ballast.[9] The hum is not only created by the design of the ballast and the way it is assembled, but it can be further amplified by the construction of the luminaire, the manner in which all components are mounted, and the acoustical traits of the location where the luminaire is used. Ballasts for higher-wattage lamps can be expected to produce more sound than those for lower wattage lamps, but this increase is somewhat diminished because fewer units are needed to achieve the required level of light.

Sometimes sound level evaluations of the luminaires are made if they are going to be used in such places as libraries, lecture halls, hospitals and similar locations where elevated noise levels become annoying.

Many ballast and luminaire manufacturers publish "sound ratings" for their products, and being aware of the problem, they are constantly improving their designs to achieve greater noise suppression.

Elenbas and Hehenkamp have noted that the noise that is produced can be divided into two categories: 1) 100 to 500 Hz and 2) 1,000 Hz and higher.[10] The low frequencies cause the structure of the iron core and its surrounding case to vibrate. However, since the ear is not very sensitive to these frequencies, they are not very bothersome, unless the units are mounted upon a very large surface area where the sound becomes considerably greater.

The noise at the higher frequencies is caused by the harmonics in the electrical current wave which cause the elements within the core to vibrate or resonate. The ear is much more sensitive to these sounds. The acoustical energy that is produced can be kept to a minimum by minimizing air gaps and joints in the core material, by winding the coil more evenly over the whole length of the core, and by using high grade, acoustical damping potting compounds to fill the ballast case.

Ballast and luminaire noises have been measured directly with a microphone and suitable amplifier, with the test unit mounted on a rubber platform and placed into a concrete box with the microphone. It was found that all of the lower frequencies were not picked up by this method.

Subsequently, sound-absorbing quiet rooms have been built and listening tests are performed to obtain relative sound levels of the units under test. With this method, many units can be checked at one time to reach a measurable level of sound, and to compensate for the differences between individual ballasts.

Another practice employs vibration pick-ups that are mounted in strategic locations on the ballast core or its housing, with the entire unit freely suspended from above. Although this method does not make it possible to measure the total noise produced, points of excessive noise amplitude can be located, which is just as important to know.

Linearity

For a ballast to provide the proper current and wattage to the lamp, it is necessary that its internal impedance does not change as the current changes. The American National Standards Institute (ANSI) Standard C82.4 defines the construction and parameters for creating a "Reference Ballast" and decrees that to meet the standard for linearity, the Reference Ballast is required to maintain an impedance value that is +/- 4% of its rated value, between the range of 50 to 115% of rated current.[11]

Deviations from this linearity criterion can result in considerable errors in measuring characteristics of those lamps that will operate at other than nominal current. In addition to the high standard maintained by the Reference Ballast, it is the goal for most commercial ballasts manufacturers who strive to achieve the 4% limit in linearity.

Even with a constant impedance and power factor, the voltage drop of most electromagnetic ballasts will increase when they are used with a lamp whose voltage is lower than nominal. This higher voltage drop across the ballast will decrease its impedance when it goes into saturation, especially if the ballast is non-linear, and this will cause more power to be delivered to the lamp.

For example, many sizes of mercury or metal halide lamps are made with a permissible range of 120 to 150 initial volts. In reactor ballast, this lamp voltage spread can create an undesirable variation of as much as -7.0 to + 8.5% in the ballast voltage. Furthermore, non-linear ballasts create distorted current waveshapes, and the relationship between the ballast current and its voltage will no longer be linear because of an accompanying increase in flux density. The distorted waveshape will usually create an unwanted increase in the current crest factor.

Extinction Voltage

Dips in the line voltage can occur whenever loads are changed or when other transient conditions exist. Properly designed and maintained distribution systems will rarely suffer dips that are greater than 10%. However, on some circuits, dips as great as 30% have been encountered. The dips are caused in many ways, such as circuit breaker interruptions, and the starting of motors, elevators and welding equipment.

These transients can be very pronounced, and without good line regulation, will be reflected back to the lamp load. A good power system is frequently called a "stiff" line, meaning that its impedance is very low and is capable of instantly supplying very high current demands, which minimizes the abrupt transient behavior.

Some marginal ballast designs will not sustain the arc within the lamp following a substantial dip in voltage. The lamp will extinguish, and then, eventually, relight.

Reactor ballasts will generally tolerate a dip of at least 15% on a well-regulated circuit. Other designs, to be discussed later in the chapter, can handle dips as great as 40%.

Studies have been made which show that not only is the magnitude of the voltage dip significant, but also its duration, as well as the rate at which it drops.[12] For ballast and lamp evaluation purposes, it has been common practice during testing, to assume a voltage dip rate of 2 to 3 volts per second.

A further interpretation of lamp and ballast stability is to actually measure the minimum steady-state line voltage at which a ballast will continue to operate a given lamp. Here again, the rate at which the voltage drops to some minimum value, becomes a factor.

Not to be overlooked is the fact that with high intensity discharge lamps, extinction, or "drop-out" can often be the function of the lamp rather than the ballast. This is particularly true as the lamp ages or otherwise deteriorates.

Temperature Rise

Most encased ballasts are designed to operate at an ambient temperature of 40°C (105°F). Ballast life will be reduced if operated at a temperature above this limit. However, in some industrial locations such as foundries, boiler rooms and heat treatment areas, the ambient temperature can reach 65°C (150°F). Special ballasts are required in these installations, otherwise the ballast life would be seriously shortened. The insulation of the core and coil assembly is made to sustain a maximum temperature limit of 165°C, which includes the actual temperature rise within the ballast, plus the maximum ambient temperature. Thus, at an ambient temperature of 65°C, the maximum permissible temperature rise of the ballast would be 100°C. Adding to the temperature rise of the ballast, is the heat generated by the lamp. For ballasts to function reliably under these conditions of extreme heat, Class "H" wire insulation which will withstand 165°C, or Class "N," good up to 200°C should be used.

For outdoor installations, when the lamps are usually burned during the cold night hours and often moving air is present, the ballasts are generally rated for 25°C ambient operation. These cooler conditions allow the use of less expensive insulating materials which reduce the ballast cost. Class "A" wire insulation, which has a maximum ballast winding temperature of 105°C, can then be used.

Heat generated in the ballast is transferred through the potting compound or resin which envelops the internal components, and is then dissipated to the surrounding air or mounting surface by conduction, convection, or radiation. Unpotted ballasts are cooled in a similar manner. It is therefore essential that the lighting enclosure in which the ballast is placed is properly ventilated. When more than one ballast is placed into a single enclosure, each should be mounted far enough apart from the others to ensure adequate heat dissipation. A well-designed luminaire can help to heat-sink the ballast which permits the use of smaller, hotter-running ballasts.

Extensive studies and temperature measurements have been made to determine the proper positioning of the ballast components within the housing, including novel ways to obtain good air circulation to remove much of the heat that is generated. Temperature measurements are not only made, but are actually specified, for some precise points on the ballast.

For example, a fluorescent lamp using Class A insulation should have a maximum ballast case temperature of 90°C and a maximum coil temperature of 105°C at normal operation.

Some ballasts are equipped with built-in automatic resetting internal thermal protection.[13] Even when a ballast is thus protected against an excessive operating temperature, it must still be well-designed to endure the normal thermal stresses expected during its rated service life. A temperature rise of just 10°C above maximum can cut the ballast life in half. This relatively small but elevated temperature can occur without even tripping the ballast's internal thermal protection device.

Fluorescent Lamp Ballast Circuits

The ballasts made to operate fluorescent lamps are not all the same. Each specific type of fluorescent lamp requires its own ballast design. Both the lamps and the ballasts are categorized according to the method by which they start and operate.

The first fluorescent lamps were instant-started. Heavy, expensive ballasts were needed to get the lamps to start and to keep them operating.

The Preheat Circuit

The preheat circuit was developed next as a way to reduce the ballast's size, weight and cost. The lamp filaments are preheated prior to lamp ignition, but a separate switching device is needed to start the lamp. Several kinds are available, but the one most commonly used is referred to as the "glow-bottle" starter. It consists of a small glass container filled with an inert gas and houses a bimetal strip attached to one of its two electrodes. When the lamp circuit is turned on, the open circuit voltage causes the gas to ionize which heats the small chamber. The bimetal then bends until it touches the second electrode, causing the lamp current to flow through the lamp filaments and simultaneously shorting out the voltage applied to the glow bottle.

The glow bottle then cools down and the bimetal quickly moves away from the second electrode. Being an inductive circuit, a voltage transient occurs. The ballast open circuit voltage, together with the starting transient pulse, provide sufficient voltage to ignite the lamp. This method of starting may take several seconds.

Lead-lag Circuit

Almost at the very onset of fluorescent lamp usage, it was found desirable to operate two lamps from the same ballast. One of the lamps is connected in series with an inductor, while the other lamp is in series with a capacitor, plus an inductor, which is still needed to help limit the current. The overall circuit will then provide a power factor close to unity.

The circuit also provides a reduction in the stroboscopic effect of the light output.

Instant Start Circuit

A better instant start circuit was developed to overcome the nuisances of pre-heat lamps such as slow-starting, blinking and starter maintenance. It is the ballast most often used to operate slimline lamps, which have just one terminal on each lamp end, rather than the customary bi-pin construction that is needed for pre-heated lamps. The slimline lamp filaments are purposely shorted by being joined at the single end terminal, so they cannot be pre-heated. Therefore, a high open circuit voltage is used to start the lamps and does so almost instantly. Cut-out lamp holders must be used for safety reasons. This method of starting tends to shorten the lamp life.

Rapid Start Circuit

The rapid start system was developed to provide the desired quick starting of the instant start circuit with the economy and size of the pre-heat ballast. A low-voltage transformer is added to the ballast to provide continuous heat for the filaments.[14] A major feature is that the open circuit voltage needed to start almost all sizes of these the lamps is less than 300 volts for a 2-lamp ballast and less than 180 volts to ground. This is achieved by grounding either the luminaire's reflector or the ballast channel cover. Starting occurs in 1-5 seconds.

The rapid start system was a major factor in acquiring fluorescent lamp acceptance into the broad commercial market.

Series Sequence Start Circuit

Series sequence-start ballasts are used to operate two-lamp circuits, but because of their unique design, they can be made smaller, lighter and consume less power than conventional lead-lag ballasts.[15] They are commonly used with slimline and high output lamps, and function in the following sequence:

1. The first lamp starts because it receives the full open circuit voltage.

2. The current flowing through this lamp is limited by an impedance network placed across the second lamp that is in series with a capacitor and the first lamp.

3. As the second lamp warms up, its impedance decreases until both lamps reach rated current. At that point, the current in the impedance network becomes negligible, so that its overall energy consumption is relatively small.

Trigger-Start Circuit

The trigger-start ballast utilizes the rapid start principle, but has a 6-volt filament heating transformer instead of the latter's 3.5 volt one.

Unlike the rapid start ballast, once the lamps are operating in a trigger start circuit, the cathode current will drop to a lower value, with a corresponding reduction in cathode power loss. The higher voltage available to the cathodes makes it possible to use this ballast with ordinary preheat lamps, up to the 30W size, and provides the desired quick starting.

Figure 6-8 shows four common fluorescent lamp circuits.

1. A simple preheat lamp circuit.

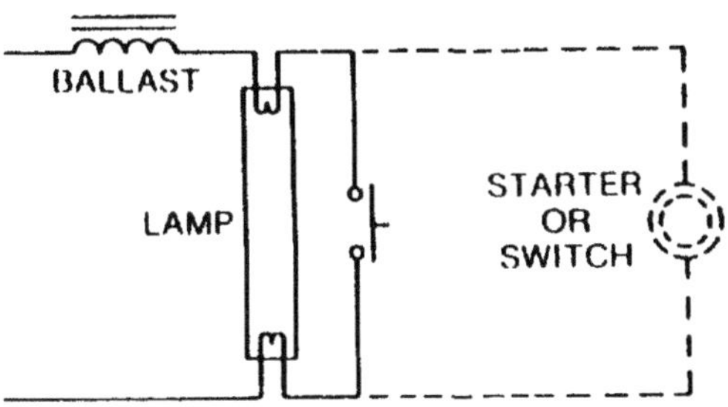

2. A two-lamp rapid start circuit.

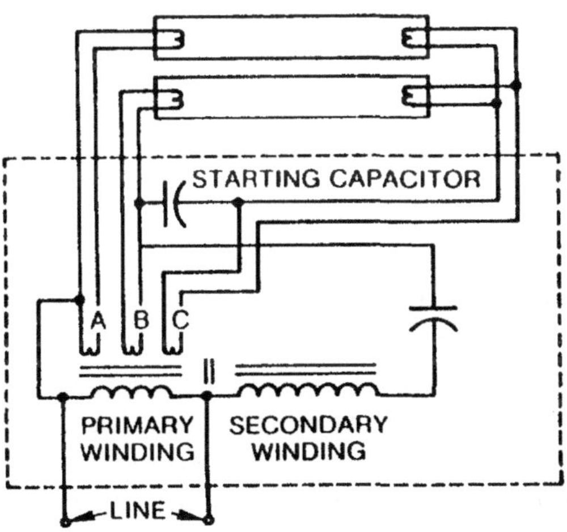

Figure 6-8. Four common fluorescent lamp ballast circuits.

(Continued)

3. A simple two-lamp instant start circuit.

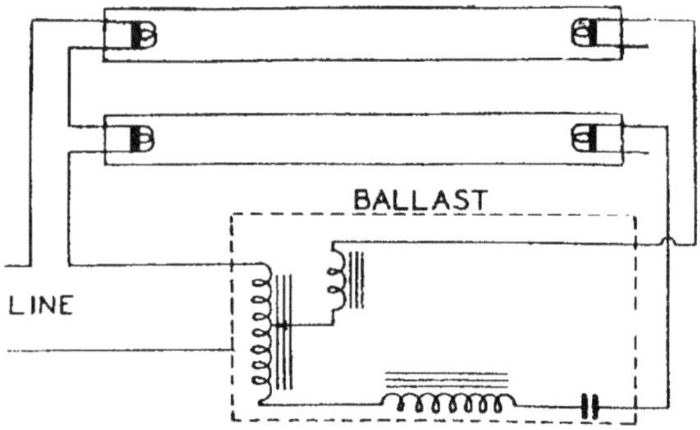

4. A two-lamp series sequence instant start circuit.

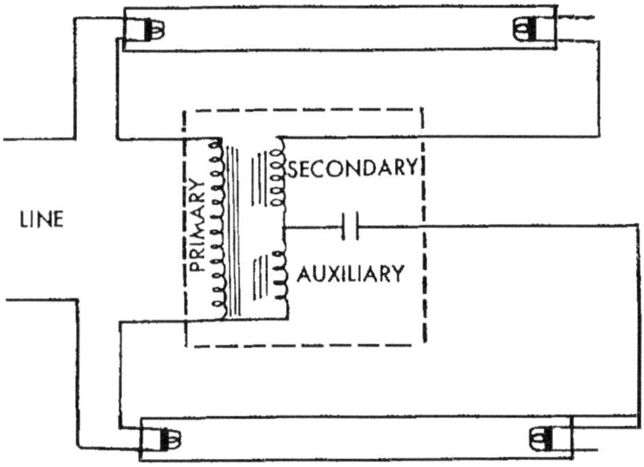

Figure 6-8. (*Continued*)

HID Lamp Ballast Circuits

A variety of ballasts are used to operate high intensity discharge lamps. Each has its own advantages and disadvantages. However, whichever ballast is selected, it must perform the basic functions of limiting the current, providing sufficient starting voltage, and sustaining the arc during each half-cycle of the lamp. It must also have a long life, have no deleterious effect on the operation of the lamp, produce a minimum of noise, and handle expected variations in the incoming voltage.

There are four basic types of electromagnetic ballasts that are used with high intensity discharge lamps, shown in Figure 6-9.

Reactor Ballast

The reactor or "choke" ballast, because of its relatively low cost, is very widely used, especially in those areas where the supply voltage is sufficient to insure reliable lamp starting. This can be 120 volts, 240 volts or even higher, depending upon the lamp's operating voltage. The reactor ballast is simple in construction, consisting of an open core of laminated steel with a coil of wire wound around one of its legs.

It is the smallest, least expensive and most efficient ballast there is. Its power losses are relatively low. Reactor ballasts can only be used where the line voltage is fairly stable and is about twice as high as the rated lamp operating voltage.

It provides for good regulation of lamp watts with variations in lamp volts, causing about a 3% change in lamp watts with an 11% change in lamp volts. Its low crest factor is close to 1.4 and it provides a high starting current that warms up the lamp rapidly. It does have several disadvantages which include the following:

1. A low power factor, but one that can be corrected with the additional expense of adding a capacitor.

2. The lamp watts can vary as much as +/- 10% with only a +/- 5% change in line voltage.

3. It consumes a large amount of current during lamp starting, so in large installations, higher rated circuit components must be used. These include circuit breakers, switches and wiring.

4. It can only tolerate line dips up to 15% for low-wattage lamps and just 10% for high-wattage lamps.

5. Because of its poor line regulation, and starting difficulties when the weather is cold, its use is limited to the more efficient and usually newer power systems, where a line voltage variation of only +/- 5% is maintained.

High Reactance Autotransformer (HX) Ballast

The high reactance ballast is also a lag-circuit design, but it consists of the coil windings of an autotransformer and a reactor wound on the

1. A reactor ballast circuit.

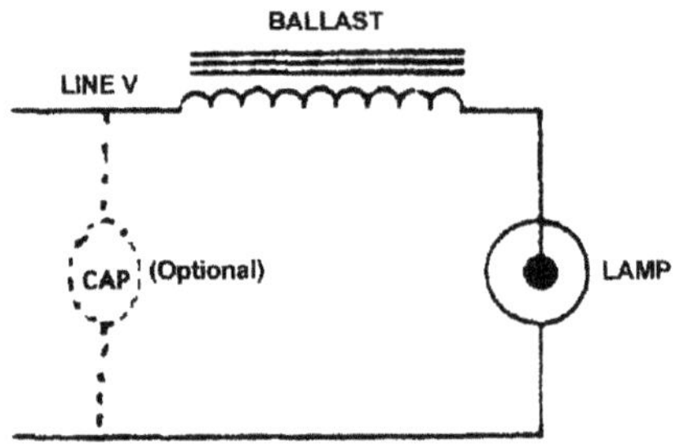

2. A regulated autotransformer ballast circuit.

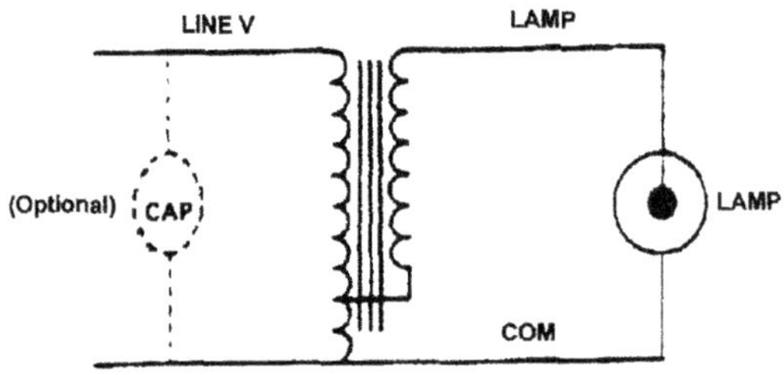

3. A regulated autotransformer ballast circuit.

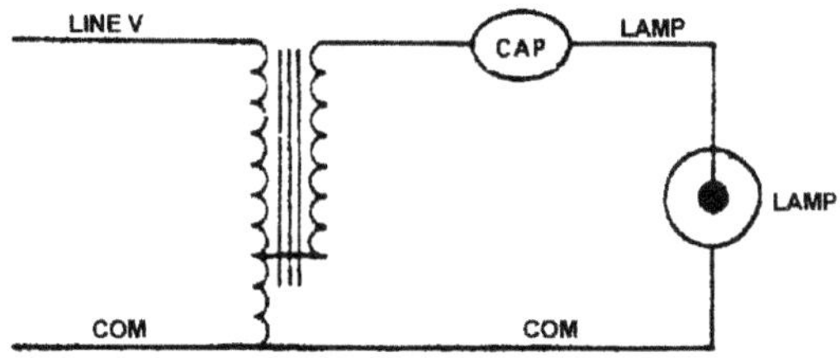

4. A regulated autotransformer ballast circuit.

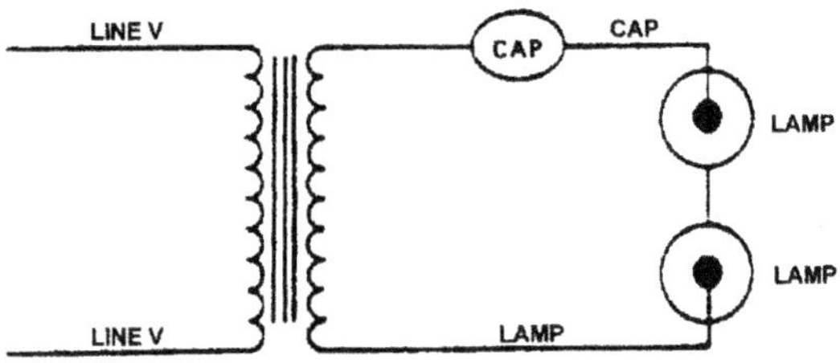

Figure 6-9. Four common HID lamp ballast circuits. Courtesy of Advance Transformer Company.

same core. It is sometimes designed with a magnetic shunt between the two windings. Its performance is very similar to the reactor, but it is generally used whenever it is necessary to increase the voltage required to start the lamp. Many of these ballasts are made with input voltage taps to operate with several line voltages.

Like the reactor ballast, a capacitor can be placed across the line voltage supply to improve the power factor.

Its disadvantages are very much the same as those of the reactor, but because of its added size, it is more costly and has greater losses than the reactor.

Constant Wattage Autotransformer (CWA) Ballast

This is a lead-circuit ballast that modifies the high reactance ballast design with a capacitor placed in series with the secondary winding to produce a high power factor. By making a portion of the primary winding common with the secondary winding, the size of core and coil assembly can be reduced. Designed to operate in magnetic saturation, the secondary winding provides good wattage regulation to the lamp and is used in installations where the line voltage fluctuation is greater than +/- 5% but less than +/- 10%.

The extent of regulation depends, in part, on the amount of primary winding that is coupled to the secondary winding.

Constant Wattage Isolated (CWI Or MWI) Ballast

This type of ballast also is made with a capacitor in series with the secondary winding and functions very much like the CWA design. It, too, is a high power factor ballast. The big difference is that its primary and secondary windings are electrically isolated from each other.

It is the kind of ballast that is generally used when two mercury vapor lamps are operated in series, to make sure that both lamp socket shells are grounded. It can be used whenever isolation is required on one-lamp ballasts. It is the most expensive of all of the electromagnetic HID ballasts.

Ballasts for Mercury Vapor Lamps

Although mercury vapor lamps are losing some of their popularity with the availability of the newer, more efficient and improved color, discharge lamps, there are still millions of installations in factories and on the roadways that still use them. Besides the high cost of fixture replacement, the fact that they last so long keeps them in demand.

The simplest way to ballast the mercury lamp, is to operate it in series with an ordinary resistor. The power efficiency in that case, is only 50%, because half of the power is consumed by the resistor. Resistor type ballasts still find use in some older, direct-current installations. In addition, it is the principle of operation for self-ballasted mercury lamps, which are used to replace incandescent lamps whose fixtures are unballasted.

The resistive ballast circuit is very sensitive to line voltage dips, and a dip of only a few percent could extinguish the lamp.

Many mercury lamps are operated on a constant-current, series regulated circuit, whereby each lamp operates from its own series transformer. The circuits are so designed that the rise in the secondary open circuit voltage is restricted when one of the lamps fails.

Bi-level and dimming circuits are also available for mercury lamps, and are often used as a means of saving energy. Mercury lamps whose bulbs are not coated are generally used in these applications, because they exhibit very little change in color when the power is reduced.

Ballasts for Metal Halide Lamps

Metal halide lamps can be considered an evolutionary change of the durable and long-lived mercury vapor lamps that have served us well for many decades. Being quite similar to mercury lamps of equivalent size, it might expected that their electrical characteristics are very much alike. However, there are some significant differences.

Metal halide lamps contain, in addition to mercury, a variety of metallic halide compounds that provide superb color rendition and high lumen efficacy. Due to the unique ionization properties of these exotic additives, two special demands are placed on the ballast.

First, a higher secondary voltage is needed to start the arc. Second, as the lamp warms up, a period of low conduction occurs in the arc plasma, and the reignition voltage needed to reestablish the arc during each half cycle can exceed the arc-sustaining voltage normally supplied by the conventional mercury lamp ballast. As a result, the metal halide lamp can extinguish during its warm-up cycle, then cool down, and again restart. Some lamps may not extinguish early in life, but will gradually reach that point in a relatively short period of burning time.

To overcome these problems, a ballast was specially developed during the 1960s for this line of HID lamps. It is called a "lead-peaked" or "peak-lead" ballast.[16]

Lead-Peaked Autotransformer Ballast

The electrical circuit of this ballast is very similar to the mercury lamp's constant wattage ballast. However, the magnetic circuit is significantly different because a section of the core that is positioned under the secondary winding is constructed with one or more open spaces within the laminations (Figure 6-10). This restricts the path of the flux and produces saturation within the core. Therefore, instead of producing a typical sine wave, the open circuit voltage waveform is sharply peaked.[17]

This high peak makes it easier to start the lamp, and generally, at the same time, will reduce the required RMS voltage. The ballast can then be made smaller and lighter.

The peaked waveform also produces a higher sustaining voltage, reduces the current off-time during each half-cycle, and reduces the current crest factor.

This type of ballast has most of the other good features of a CWA ballast, such as allowing just a 10% change in lamp watts with a 10% change in line voltage, a low line starting current, a high power factor and tolerates reasonable line voltage fluctuations.

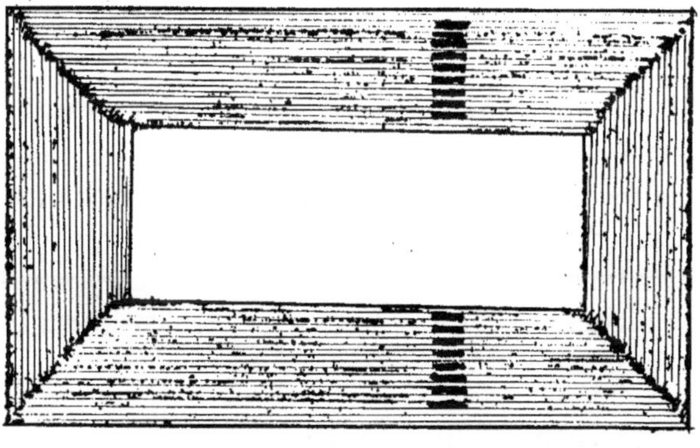

Figure 6-10. Cross-section showing openings placed in the core for a lead-peaked metal halide lamp ballast (from Cohen, 1965). Courtesy of the Illuminating Engineering Society.

Significant improvements in lamp design have reduced the restriking voltage and arc-sustaining current, and now many metal halide lamps can be operated on ordinary mercury lamp ballasts. Their use is restricted by the availability of sufficient line voltage, proper line voltage regulation and reasonable limits of line voltage dips and transients. In some cases, an external ignitor is required to aid starting.

Furthermore, there are new ballast circuit designs for low-wattage lamps that now incorporate an internal pulse-ignitor. This makes possible the elimination of the starting electrode as well as other alterations in the arc tube design and permits an increase in the gas fill pressure. All of these modifications help to improve the lamp lumen efficacy and lumen maintenance.

New linear reactor circuits with high saturation levels have been developed. They operate with the same starting current as standard CWA ballasts, but can be made at a lower cost and with less power loss. These linear ballasts can operate 320W to 400W metal halide lamps with only 277 volts input. They are generally made with built-in ignitors to insure reliable starting.

There are now also available regulated-lag circuit ballasts that are used with metal halide lamps where good wattage regulation throughout life is required. However, these ballasts are relatively expensive and have a high power loss.

Ballasts for HPS Lamps

Unlike mercury vapor lamps, which operate at almost constant voltage over their lifetime, high pressure sodium lamps demands on maintaining good wattage regulation are much more rigorous. One of the problems is that HPS lamp voltage rises substantially during lamp life, primarily because of the gradual loss of sodium within the arc tube. The sodium, as a hot and highly reactive element, readily combines with other materials in the arc tube. The ratio of the sodium-mercury amalgam then changes to a higher proportion of mercury, which increases the voltage.

Other Conditions that Affect the Lamp Voltage

In the mercury vapor lamp, the mercury vaporizes completely, which keeps the vapor pressure constant during its life. Most HPS lamps, on the other hand, contain an excess amount of the sodium-mercury amalgam, which can be either vaporizing or condensing as the arc tube temperature fluctuates. As a result, the arc tube gas pressure, and consequently, the lamp voltage changes.

Over time, vaporized tungsten, barium and aluminum will condense at the ends of the arc tube. This deposit on the inside wall builds up, absorbing heat from the arc stream, causing the arc tube temperature to rise, which increases the vapor pressure and voltage.

Radiant energy can be reflected back to the lamp from the luminaire to increase the arc tube cold spot temperature and cause the lamp voltage to rise.

Furthermore, variations in line voltage, lamp and ballast manufacturing tolerances produce significant variations in lamp wattage right from the onset. All of these variations manifest themselves in broad changes in lamp color and light output.

The Development of Performance Curves

Shortly after the lamp's introduction, it became necessary for lamp and ballast manufacturers to develop performance curves to show lighting designers and other end users what to expect to get from their lighting installation, over time, for each HPS lamp design.[18]

The task of generating these curves was accomplished through the efforts of ANSI. Lamps that were seasoned for 100 hours were measured for their voltage and wattage on a specially-built, high-performance linear ballast generally referred to as a "Standard Reactor."

Parallelograms, called "ballast trapezoid diagrams," were developed which recommended to the ballast manufacturers, the preferred operating limits of lamp voltage and lamp wattage for each lamp design.

The total recommended voltage range as shown on the trapezoid diagram lies between the minimum and maximum voltage lines.

Because lamp light output is directly related to the lamp wattage, it is also necessary to show in the trapezoids, the recommended range of lamp wattage spread.

The trapezoid diagram that was developed for the 400W HPS lamp is shown in Figure 6-11.

Exceeding the recommended maximum wattage limit will shorten the lamp life, reduce long term lumen maintenance and may cause the lamp to extinguish during operation. Operating below the minimum wattage limit will result in low lumen output, unduly extend the lamp's warm-up time, and may not even permit the lamp to start.

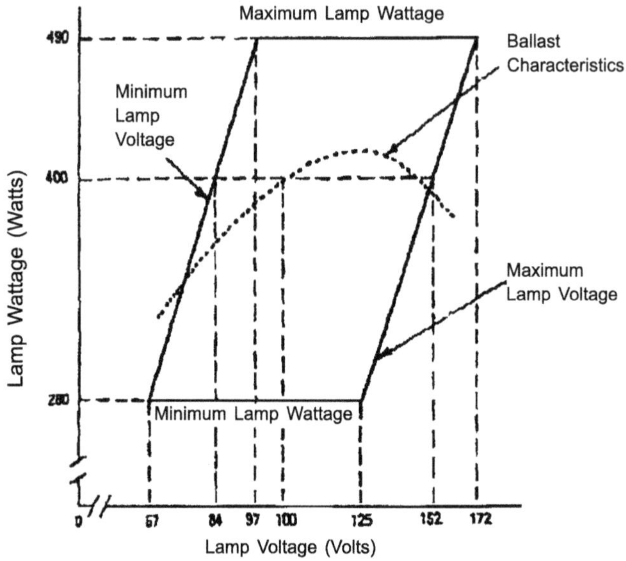

Figure 6-11. The ANSI trapezoid for a 400W HPS lamp (from ANSI Standard C78.1350-1990). Courtesy of the American National Standards Institute.

The Ballast Characteristic

Acceptable lamps coming off of the production line can vary as much as +/- 15% in voltage, as measured at their rated wattage center. Additional provisions must be made in the ballasts to allow the lamps to continue to operate satisfactorily as the lamp voltage rises during life. The curve that the stabilized 100-hour lamps generate is called the "ballast characteristic."

This wattage-vs.-voltage curve shows the path the lamp's wattage will change as the voltage rises, while the input voltage is held constant. All lamps of the same design will operate on this curve, when operated on this ballast. If the input voltage changes, as is normally expected to happen in service, the position of the entire ballast characteristic curve will shift upward or downward, bringing about a corresponding change in lamp wattage.

Ballasts other than the linear reactor ballast will have their own ballast characteristic (curve). Many provide better input voltage regulation than the linear reactor ballast, and will reduce the variation in lamp wattage significantly, over the useful life of the lamp.[19] Figure 6-12 shows the ballast characteristics produced when 400W lamps are operated at the upper and the lower limits of the line voltage. It also shows how the lamp wattage spread differs with the type of ballast used. Ballast characteristic curves generated by HPS lamps of other wattages are very similar in shape.

The path of the ballast characteristic comes to an abrupt end, when the lamp voltage has risen so high that the lamp extinguishes. This is called the "drop-out" point, and is recognized as where the lamp has reached the end of its life.

Reignition Voltage

To further explain "drop-out," it is necessary to examine the lamp's instantaneous reignition voltage.

In Figure 6-13, when voltage is plotted against time, it can be seen that as the instantaneous lamp voltage goes out at point, "t," there must

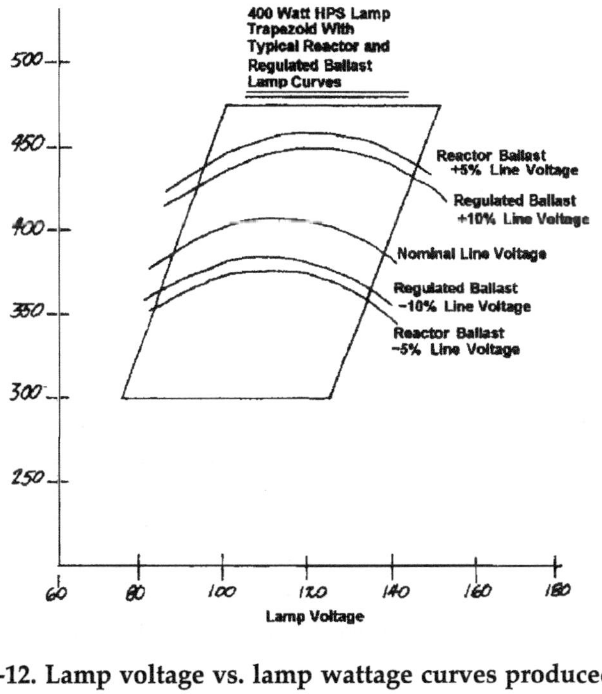

Figure 6-12. Lamp voltage vs. lamp wattage curves produced by 400W HPS lamps operating on a reactor and a regulated ballast. Courtesy of Thomas & Betts Lighting Division.

be sufficient instantaneous open circuit voltage available to reignite it. Fortunately, in most ballast designs, the open circuit RMS voltage is generally more than twice the lamp operating voltage, so reignition is not a problem, at least not until the lamp nears the end of its life.

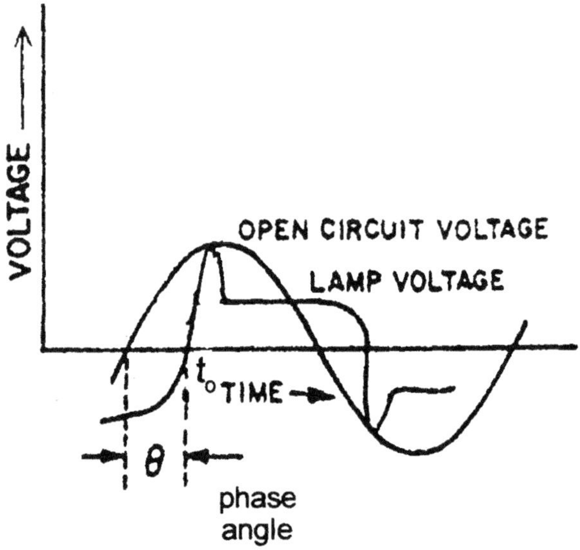

Figure 6-13. The relationship of the instantaneous open circuit voltage to the lamp voltage in an HPS lamp.

A high starting current would seem desirable to warm the lamp rapidly so that it will reach stabilization quickly; as it flows through a cold electrode, however, it will cause the tungsten to sputter and vaporize more readily.

The Starting Pulse

The electrical operating characteristics of high pressure sodium lamps are very similar to other high intensity discharge lamps, but the lamps require much higher voltage to start. Starting assistance is provided with an auxiliary electronic device that places a high voltage pulse on each cycle of the open circuit waveform until starting is achieved.

The effectiveness of the starting pulse depends not only on its shape, magnitude and duration, but also on its precise location near the peak of the open circuit voltage waveform,[20] as shown in Figure 6-14. All of the pulse parameters must be carefully controlled to achieve sufficient ionization of the xenon starting gas and to maintain this ionization until the current wave passes through zero. When the lamp starts, the pulse shuts off.

Many patents have been granted for the great variety of electronic ignitor circuits that have been designed to generate pulses of the desired

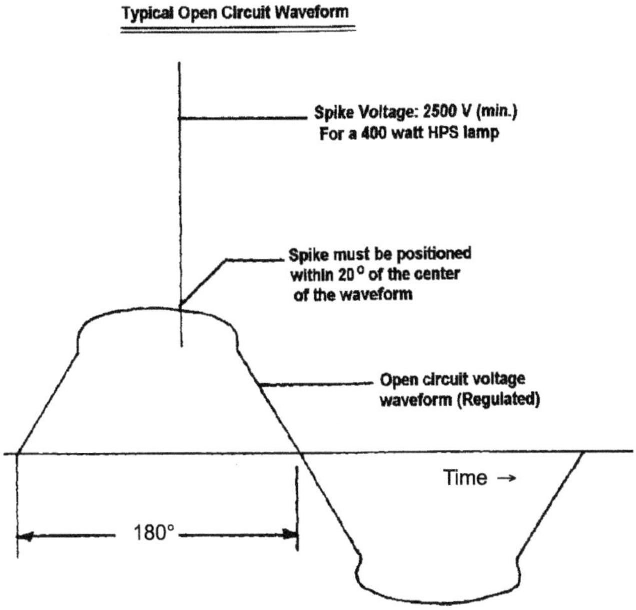

Figure 6-14. Proper location of spike on open circuit voltage wave (from Shelby, 1974). Courtesy of Thomas & Betts.

shape, amplitude, time duration and precise location on the secondary voltage waveform. These starting devices are either built into the ballast or are separate units that can be mounted externally, and even remotely.

HPS Ballast Circuits

The three general types of ballasts most commonly used to operate HPS lamps are the lag circuit, lead circuit and magnetic-regulated ballasts.

The lag circuit ballast is the same as the ordinary HID reactor ballast discussed earlier in this chapter. It is sometimes used with an autotransformer to accommodate low line voltages. As with other choke ballasts, it is small in size, relatively inexpensive and has a low power loss. Its high starting current can produce a faster lamp warm-up than other ballasts.

The lead circuit ballast often includes an autotransformer. As with other HID ballasts, by making a portion of the primary winding common to the secondary winding, the ballast size can be kept small. This type of ballast has a low starting current and high power factor.

The magnetically-regulated ballast that was developed for HPS lamps is essentially a voltage-regulating, isolation transformer with the primary and secondary windings wound around two legs of a common core. A capacitor is connected to a separate third winding, and this circuit adjusts the magnetic flux within the core when either the primary or secondary voltage changes. This design provides the best regulation of lamp wattage when either the line voltage or the lamp voltage vary. Such ballasts are large, costly and have a substantial power loss. On the other hand, they have a high power factor and produce a very low line current.

Considerable work has been done in the development of electronic ballasts in order to provide a more constant wattage as the lamp voltage changes, or as the line voltage changes, or to sense and compensate for any change in lamp voltage as soon as it occurs. To reduce their size and weight, ballasts that operate at high frequency have been made which can provide current waveforms that are either pulsed or sinusoidal.[21] Other high frequency ballasts can provide instant re-ignition if the lamp extinguishes as a result of a power interruption.

In spite of their good features, the fact is that electronic ballasts are still relatively expensive, are prone to damage from line transients and to deterioration from the intense heat found in some fixtures. Most of the ballasts that are still used to stabilize HPS lamps are the conventional magnetic types.

Satisfactory lamp performance throughout life can be expected with any of the HPS ballast designs so long as they can operate the lamp close to its rated wattage.

Electronic Ballasts for Fluorescent Lamps

It was recognized in the late 1960s that the efficacy of fluorescent lamps could be substantially increased by operating them at frequencies above 60 Hz. Ballasts designed for high frequency operation use small electrical components making them smaller and lighter in weight than conventional electromagnetic ballasts.

At low frequencies, relatively heavy magnetic inductors, usually in combination with a capacitor, are required to achieve satisfactory lamp performance. At high frequencies, low-loss capacitors and ferrite inductors can be used without an adverse effect upon the waveshape of the lamp's current and voltage.

With the increased use of semi-conductor circuits in electrical products, many ballast manufacturers were quick to come up with a great variety of circuits for operating fluorescent lamps. It took many years of testing and careful evaluation to determine the effect of high frequency electronic circuits on fluorescent lamp life, lumen maintenance, color shift and end blackening.[22] But now these electronic ballasts are commonplace and they comprise a substantial percentage of the market.

Other modifications include electronic starters for preheat fluorescent lamps,[23] electronic dimming circuits[24] and programmable controls.[25]

Rectifiers are used to convert the 60 Hz input power to low-voltage DC, which is then converted to high frequency with transistors. Low-voltage DC can be obtained directly from batteries which makes it possible to provide fluorescent lighting in lanterns, trains, buses and airplanes.

Other benefits provided by electronic ballasts include the stabilization of the lamp current over broad variations in ballast voltage and, at high frequencies, the ballast noise is virtually inaudible.

As electronic ballasts became more available, a newer development combines the magnetic core style with the electronic circuit. It is called the "hybrid" ballast. It makes possible, simplifies or improves such applications as dimming, lamp wattage control and multiple on-off control.[26,27] It is lighter in weight than the electro-magnetic ballast and it can be made at a lower cost than the totally electronic ballast. It was essentially developed to meet new federal regulations which require new commercial buildings to reduce the energy consumed by their lighting installations. Many of these new ballasts can reduce the energy consumption by 25% with little or no reduction in the light output. Furthermore, they can also function with electronic controllers that will provide two-level (night/day) lighting automatically.

With the introduction of compact design fluorescent lamps, electronic ballasts were obliged to be made even smaller, so that the lamp and ballast could be sold as an integral unit.[28]

Producing electronic ballasts to operate high intensity discharge lamps took much longer in coming. Problems were encountered in trying to achieve lamp stability at high frequencies, primarily because of a phenomenon called "acoustic resonance." After much investigation, by several lamp development laboratories, it became clear that the resonance could be diminished by reducing the DC ripple created in the rectifier circuit, or by modulating the high frequency waveform.[29]

Other improvements were soon added. These include igniting the lamps with electronic starting devices[30] and providing pulsed lamp current operation.[31]

References

1 Lin, K.C. and Ellis, B.B., "Progress of Core Technology in Transformer Manufacturing," *Insulation/Circuits,* 1977.

2 Hehre, F.W. and Harness, G.T. *Electric Circuits and Machinery,* Volume 2. New York: John Wiley and Sons, 1942.

3 Martt, E.C. and Smith, R.J., "Mercury Lamp Current Crest Factor and Its Effect on Lumen Maintenance," *Illuminating Engineering,* 1958.

4 Wide Lite Corporation, "Effect of Lamp Current Crest Factor on Lamp Lumen Maintenance," July 1972.

5 Smith,R.J. and Matz, W.C., "Modified Current Wave Shape and Starting Voltage Limits For Mercury Lamps," *Illuminating Engineering,* 1962.

6 Noel, E.B. and Martt, E.C., 1956. "Effect of Operating Variables on Mercury Lamp Performance," *Illuminating Engineering,* 1956.

7 Scholz, C.F., "Characteristics of Ballast Regulations For High Intensity Discharge Lamps," *Illuminating Engineering,* 1974.

8 Varnum, V., "What Is Power Factor?," NAPLC Sales Bulletin, 1965.

9 Certified Ballast Manufacturers Bulletin, 1969. *Why Do Some Ballasts Hum?*

10 Elenbas, W. and Hehenkamp, T., *High Pressure Mercury Vapor Lamps and Their Application.* Philips Technical Library, 1965.

11 American National Standards Institute. C82.4-1992: "Standard For Ballasts For High Intensity Discharge and Low-Pressure Sodium Lamps."

12 Burns, C.H., 1963. "Effect of Ballast Circuitry and Line Voltage Variations on Mercury Lamp Stability," *Illuminating Engineering,* 1963.

13 MagneTek Company Lighting Products Catalog, 1995.

14 Lemmers, A.E. and Brooks, W.W., "New Fluorescent Lamp and Ballast Design For Rapid Starting," *Illuminating Engineering*, 1952.
15 Hinman, D., Keiffer, R.L. and Strecker, C.E., "Fluorescent Lamp Performance on Sequence-Start Ballasts," *Illuminating Engineering*, 1951.
16 Cohen, S., "Ballasting The New Metallic-Additive Discharge Lamps," *Illuminating Engineering*, 1965.
17 Waymouth. J.F. *Electric Discharge Lamps*. Massachusetts: Mit Press, 1971.
18 Unglert, M.C. and Kane, R.M., "The Interaction of High Pressure Sodium Lamps, Ballasts and Luminaires," *Lighting Design & Application*, 10-11:P.43-47, 1980.
19 General Electric Company Ballast Bulletin, 1974. *Ballasts For High Intensity Discharge Lamps.*
20 Shelby, B.L., "High Pressure Sodium Lighting: An Overview," *Electrical Consultant Magazine*, September 1974:P.26.
21 Degroot, J.J. and Van Vliet, J.A.J.M. *The High Pressure Sodium Lamp*. Philips Technical Library, 1986.
22 Hammer, E.E., "Fluorescent Lamp Operating Characteristics at High Frequency," *Journal of the IES*, October 1984.
23 Gyoten, G., Ito, K. and Yoshikawa, N. "Development of Electronic Starters For Fluorescent Lamps," *Journal of the IES*, 1995, 24-2:P.86.
24 Okude, A., Ueoka, A. et al., "Development of An Electronic Dimming Ballast For Fluorescent Lamps," *Journal of the IES*, 1992, 21-1:P.15.
25 Christensen, B., "Electronic Ballast Improves Fluorescent Lamp Efficiency," *Lighting Design & Application*, 20-1:P.13, 1990.
26 Garbowicz, G., "An Improved Hybrid Fluorescent Lamp Ballast," *Journal of the IES*, 21-2:P.42, 1992.
27 Garbowicz, G., "Hybrid Ballasts II: High Efficiency Type Ballast For 32W T8 & 34W T12 Lamp Systems," *Journal of the IES*, 1994.
28 Rasch, E., "Properties and Performance of Compact Fluorescent Lamps With Integral HF Electronic Ballasts," *Lighting Design & Application*, 1987.
29 Ibid, Degroot, J.J. and Van Vliet, J.A.J.M., 1986.
30 Pabst, W. and Klein, D., 1992. "Igniting High Pressure Lamps With Electronic Ballasts," *Journal of the IES*, 1992.
31 Nishimura, H. et al, "A New Electronic Ballast For HID Lamps," *Journal of the IES*, 1988.

Chapter 7

Fluorescent Lamps

by Dr. Dieter Hofmann and Erhard Rasch

Principle and Importance

Fluorescent lamps are low-pressure gas discharge lamps. In the discharge, ultraviolet (UV) radiation is generated. A fluorescent material, commonly called "phosphor," converts the UV into visible radiation—i.e., into light.

Both processes, the generation and the conversion of UV, are highly efficient. Therefore, fluorescent lamps are among the sources for white light with the highest efficacy (lumens per watt, or LPW).

There are other lamps that emit more lumens for the same wattage, the most important of which are the sodium high pressure and low pressure lamps (see Figure 7-1). But instead of white light, high pressure sodium lamps typically produce a yellowish light with reduced color rendition, while low pressure sodium lamps emit a monochromatic yellow light which makes all things appear with the same color ("none"); only brightness differences are detectable.

Because of their high efficacy, good color rendition and other advantages, fluorescent lamps and their descendants are nowadays the most important sources for producing light. More than 50% of all electric lighting applications use these lamps.

Construction and Theory of Operation

A standard fluorescent lamp consists of a glass tube with a phosphor layer on its inner surface. The tube contains a special gas filling as a discharge carrier, and two electrodes. Figure 7-2 shows one of the elec-

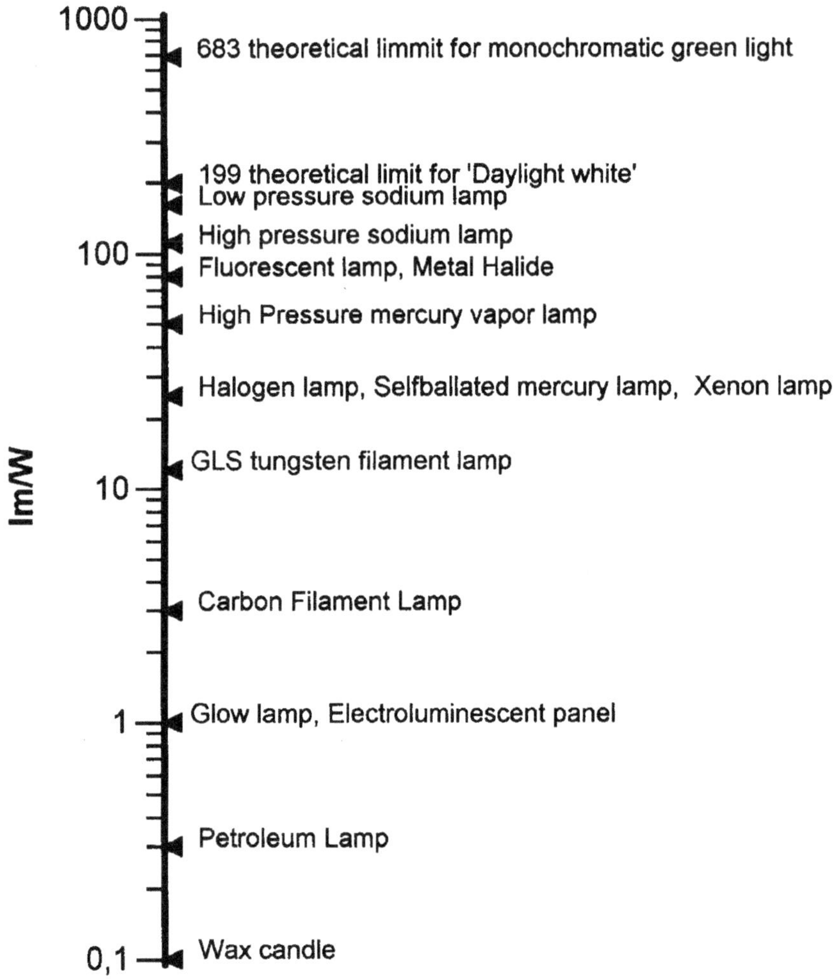

Figure 7-1. Typical efficacies for several lamp families.

trodes. The - and + indicate that a voltage is applied to the lamp. The electrode is assumed to be heated up to a temperature that allows electrons to be emitted into the space in front of the electrode. The electric field accelerates the electrons. They gain kinetic energy in the field.

Up to a certain energy, the electrons are elastically scattered by the other gas particles. But as soon as their energy is high enough to convert a gas atom from the ground state to a higher (electronic) energy level, they may transfer their energy to the atom. This is called

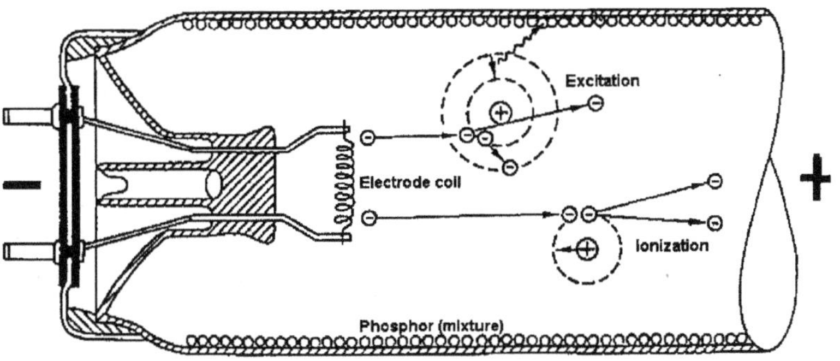

Figure 7-2. Basic processes in a gas discharge: excitation and ionization.

the "excitation" process.

The excited atom will return to its ground level within a very short time, emitting the energy difference as a UV photon. This photon, when absorbed by a phosphor particle, will cause the phosphor to emit light.

Besides the excitation, there is another important basic process: the "ionization" of atoms. Not all electrons lose their energy by exciting atoms. There are some, that gain even higher energy, so high that they may not only excite an atom, but may tear an electron completely from the atom. The result is an ion, or positively charged atomic remnant, and another free electron.

This ionization process is necessary for two reasons. First, the positive charged ions "neutralize" the negative space charge of the electrons that would create an electric field counter to the one applied from outside. Second, the additional free electrons are needed to replace those that are lost by recombination processes or by impact with the wall. Without the ionization processes, only a small electric current could flow through the lamp, much smaller than needed to produce as much light as is usually generated by a fluorescent lamp.

History

"Prenatal" Phase

The main processes, gas discharge and fluorescence, are "prehistoric": Lightning flashes are gigantic gas discharge phenomena, and some natural minerals fluoresce without any human aid, Willemite

($ZnSiO_4$) being one of the best known.

In ancient China, it was known that some sulfur compounds glow at night after being irradiated by sunlight. But this effect is not strictly fluorescence, it is phosphorescence, as the luminescence (the generic term for both) *after* excitation is named. Fluorescence, the emission of light *during* excitation, became known only when experiments with invisible exciting radiation started, or since spectral analysis was applied to detect the fluorescence by showing the different spectral distributions of exciting and emitting radiation.

Becquerel[1], experimenting with gas discharge vessels, saw that some substances lighted up brightly when being exposed to the discharge. He is considered to be the inventor of fluorescence. Among others, Edison investigated fluorescence phenomena. In 1896, he suggested[2] the use of $CaWO_4$, because it is a very stable phosphor. In 1923, Risler[3] and Claude[4] reported about the lighting of several materials in a neon discharge tube. In 1924, Ruttenauer and Wendt[5,6] found that they could improve the fluorescence of the well-known Willemite by adding manganese.

Although experts noticed in the first half of the 1930s that the combination of low pressure discharge and fluorescence could be the basis for a revolution in general lighting technology, the application of this new principle of illumination in the following years was restricted to what we call today "advertising" or "neon tubes," with and without phosphor, possibly of gigantic dimensions. The illumination shown by Osram during the World Exhibition in 1937 in Paris was of the same type. Nevertheless, this illumination impressed the International Jury so much that it was awarded with the golden medal in the illumination class.[7]

The Birth of The Fluorescent Lamp

In 1936 and '37, several development groups worked on lamps, with technologies very close to our recent fluorescent lamps.[8,9,10,11,12,13] However, their lamps were more of experimental character than marketable products.

An important step toward the new lighting era was made in the United States by the General Electric Company, the successor to the Edison Company. In 1938, it introduced a whole family of the new lamps with 15W, 20W and 30W power consumption and seven different light colors.[14,15] Similar to Edison, whose success was mainly based on offering not only the bare incandescent lamp but also all that was needed to operate it, now, 60 years later, the new lighting system was also completed by the accessories for the lamps—ballasts and a magnetic starter. The new lamps did not need high-voltage transformers like most of the previous neon tubes; they could be operated on line voltage. The new lighting system was shown at the World Exhibitions in New York and San Francisco that opened on April 1, 1938.[16]

Another fact comparable to Edison's incandescent lamps is remarkable: Like the screw-in base introduced by Edison that is still used today, the standard 2-pin-base of the present fluorescent lamps dates back to those first commercial lamps.

Main Development Steps

Although their technology was very similar to that of our present lamps, the performance of the first fluorescent lamps was far from that of a modern fluorescent lamp. Lamp life was not much longer than that of an incandescent lamp. The luminous efficacy was about 30 LPW, less than one-third of what we can produce today. Nevertheless, this was still three times more efficient than a contemporary standard incandescent lamp, a value that made the new lighting system very attractive to users. However, it was less attractive to the utility companies, who feared losses in the sale of electricity, particularly since the percentage of electricity used for lighting was higher at that time than it is today.

Lamp Life

The life of a fluorescent lamp is limited by the electrode systems. The first electrodes were "coiled coils" similar to those of incandescent lamps, but the emission material, needed to increase the thermionic emission of electrons, crumbled. "Triple coils" (in reality "coiled coils" bearing a very fine overwind wire), invented by GE in 1941, avoided this problem. "Stick-type" coils use the same principle and need fewer production steps (see Figure 7-3). The two latter types are still in use today.

The first emission material was the same as that then used in high vacuum electron tubes, based on barium oxide. But the original material was not resistant against the much higher currents in a fluorescent lamp and against the ion bombardment. Additives like strontium oxide and calcium oxide, and later on zirconia or zirconium, were found to improve the behavior of the material. Within the first 10 years of the devel-

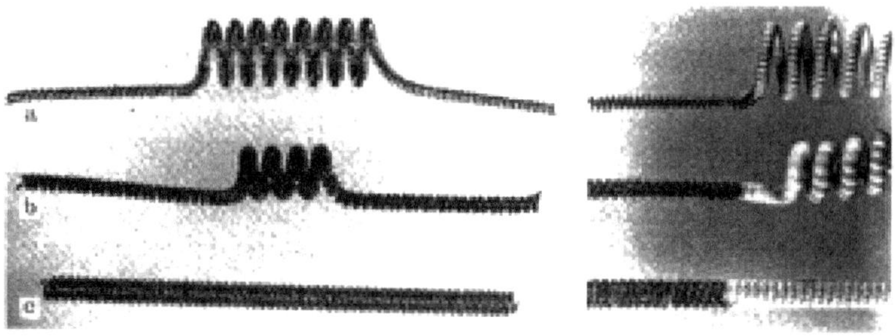

Figure 7-3. Electrode coils for fluorescent lamps: "coiled coil" (a), "triple-coil" (b) and stick-type coil (c). At right: coated with emission material.

opment effort, lamp lives up to 6,000 hours were realized.

The life of a lamp is not only determined by the electrodes. The current limiting circuitry of the ballast, is an additional important factor, and the loss of luminous efficacy during the operating time has to be considered too. Usually, not only a lamp that has failed (ceased to emit light), but also one that emits less than 70% of the initial lumens is considered to be at the end of its life.

Luminous Efficacy

Since 1930, the efficacy of fluorescent lamps has increased dramatically from about 30 LPW to now 100 LPW and more (see Figure 7-4). This improvement was achieved by intensive development work with the target to optimize the known phosphors and the technique of their application. Thus, during the 1940s, an increase up to about 50 LPW was achieved essentially by improving the silicate phosphors. Among these phosphors was the above mentioned $ZnSiO_4$, manganese activated, with about 100 LPW, but with a strict green emission. Other phosphors or phosphor blends had to be used to produce white light, suitable for general lighting. Many recipes were developed to mix phosphors of different emission spectra, and as their spectral distributions were various and relatively broad bands, it was no problem to get comfortable light, containing all wavelengths from blue to red. These phosphor mixtures or their components have been used for many years in so-called "deluxe" fluorescent lamps.

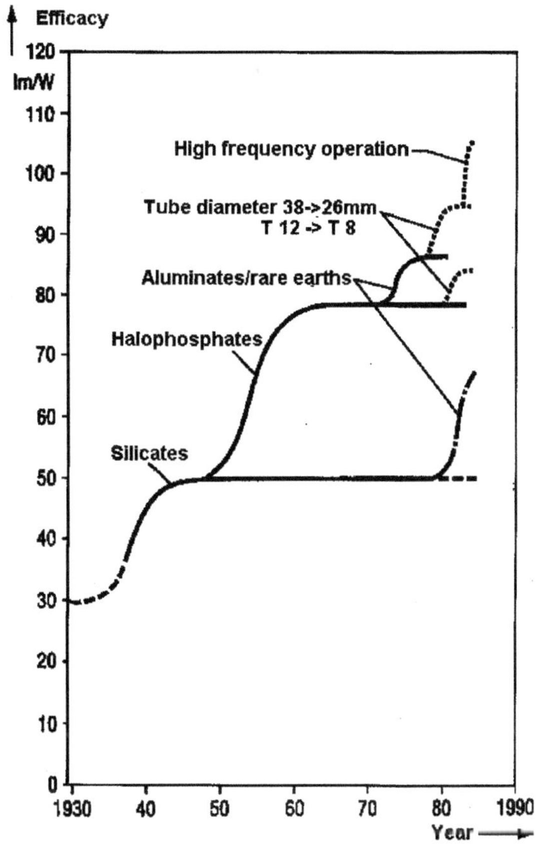

Figure 7-4. The efficacy of typical fluorescent lamps, improvements from 1930 to 1990.

But the width of the spectrum was not always an advantage. The lamps generated much light at the short- and long-wavelength ends of the visible region, where the human eye is very insensitive. In other words, they "wasted" energy, and, therefore, with those phosphors an efficacy over 50 LPW was not possible.

Halophosphate

If the introduction of fluorescent lamps was one revolutionary act in the 1930s, the other came in the 1950s. This was the invention of the halophosphate phosphors in the laboratories of the British General Electric Company[17] and its subsequent introduction.

Halophosphate is a derivative of apatite, $Ca_5(PO_4)_3F$, where some calcium is replaced by antimony and manganese, and some fluorine by chlorine and oxygen. The special property of this material is the simultaneous emission of two fluorescence bands from one single phosphor—one in the blue, the other in the yellow region. These two bands add up to form white light.

The intensity ratio of the two bands can be widely changed by varying the concentration of the activator additives, especially the manganese (see Figure 7-5). Therefore, it is possible to prepare all "white colors," that are desired for different general lighting applications, from warm white to daylight, from one basic material.

Figure 7-6 shows that (with the exception of the blue L129) the halophosphate colors are close to the Planckian (black body locus) curve and close to Judd's lines for 3000K, warm white (WW); 4000K, cool white (CW); and 6500K, daylight (D).

Therefore, the emitted light by definition may be called "white light."

The emission spectra of the halophosphates decrease rapidly beyond 630 nm. This is both an advantage and a disadvantage. The advantage is that less ineffective long-wave radiation is wasted. The result is a much higher efficacy of the halophosphate lamps compared to the

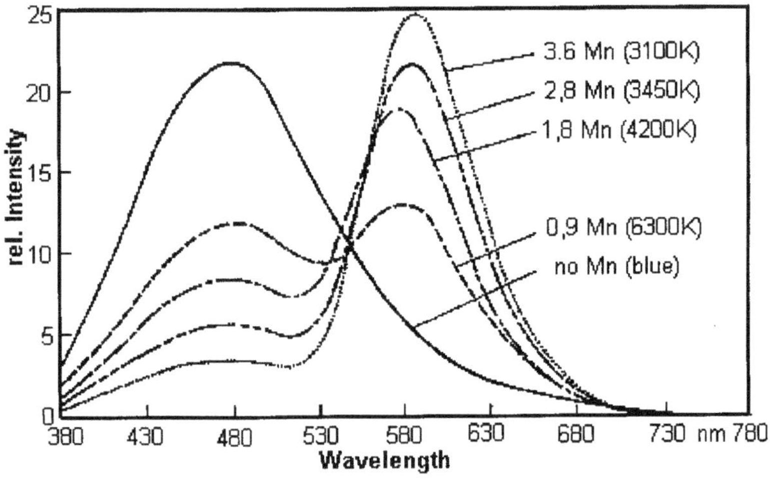

Figure 7-5. Emission spectra of halophosphates with different activation.

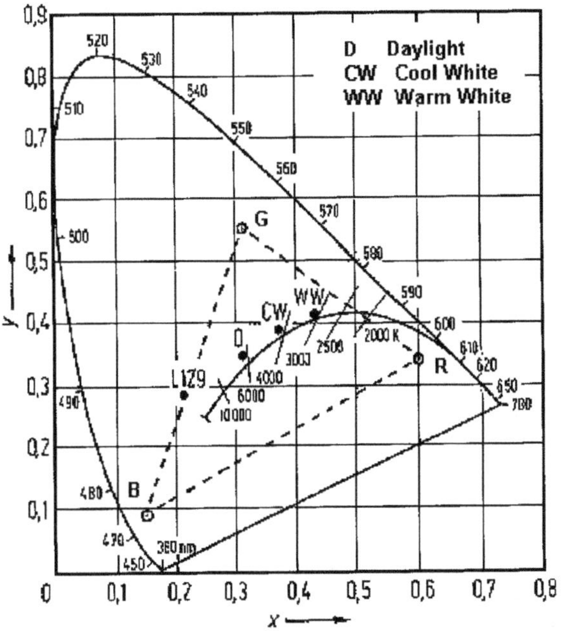

Figure 7-6. Color coordinates of halophosphates and of the red, green and blue tricolor phosphors (B, G, R).

deluxe lamps (nearly 80 LPW instead of 50 LPW). The disadvantage is that the lack of red makes the light less comfortable.

Tricolor Lamps

Up to the end of the 1960s, a customer had to choose between deluxe lamps with excellent color rendition but low luminous efficacy, and halophosphate lamps with high efficacy but reduced color rendition. This situation changed with the realization of the "tricolor principle," which resulted in lamps with good color rendition and even higher efficacy than all other lamps known at that time. It was another revolutionary step in the history of fluorescent lamps. However, a customer was confronted with another decision—whether or not to pay the higher price for the new lamps.

The basics of the new principle were computerized calculations to investigate the relationship between the spectral distribution of a light source, and the color rendering index (CRI) on the basis of the definitions that were worked out some years before.[18,19] These investigations were made independently in different places.[20,21,22,23] The results were that, even with a light spectrum that consists of only three monochromatic lines, a high CRI could be realized. Detailed calculations revealed that among all combinations of three wavelengths, a maximum value, CRI = 85, is obtained by choosing optimally 450 nm (blue), 540 nm (green), and 610 nm (red).

Thornton at Westinghouse was the first who demonstrated a lamp made with a mixture of a blue, a green and a red phosphor, emitting close to the optimal wavelengths.[20] He used Y_2O_3:Eu for the red (YOE), that was known for several years by the Osram laboratories.[24,25] It is an efficient and stable red phosphor that is still the standard red for tricolor lamps. However, Thornton's green and blue phosphors were not ideal. The zincsilicate green was especially unstable. This is unfortunate, as zincsilicate is a low-cost phosphor and the spectral distribution of its fluorescence is closer to desired values than expensive phosphors developed later. But at that time, the result was a lamp with the expected high CRI, but with a luminous efficacy and lumen maintenance poorer than that of a halophosphate lamp. Under these circumstances, the higher price of the lamp, due to the expensive red phosphor, seemed unjustifiable.

A new era began with the development of two new phosphors at the Philips laboratories: The green-emitting cerium-magnesium aluminate, terbium-activated ("CAT")[26] and the blue-emitting barium-magnesium aluminate ("BAM"), Eu2+ activated[27] phosphors, both with high-quantum yield and excellent stability, even in highly loaded lamps. These two phosphors, together with the above mentioned red-emitting YOE, provided the basis for a new lamp generation.[28] Typical spectra of a halophosphate, a tricolor, and a new generation of deluxe lamps are reproduced in Figures 7-7, 7-8 and 7-9.

The "Second Lamp Optimizing" Step: Energy-saving

The first investigations to find optimum lamp performance data were made in the middle of the 1930s, even before the fluorescent lamp was really born.[29] At that time, it was found that the discharge vessel should have a diameter of about one to two inches and, depending on

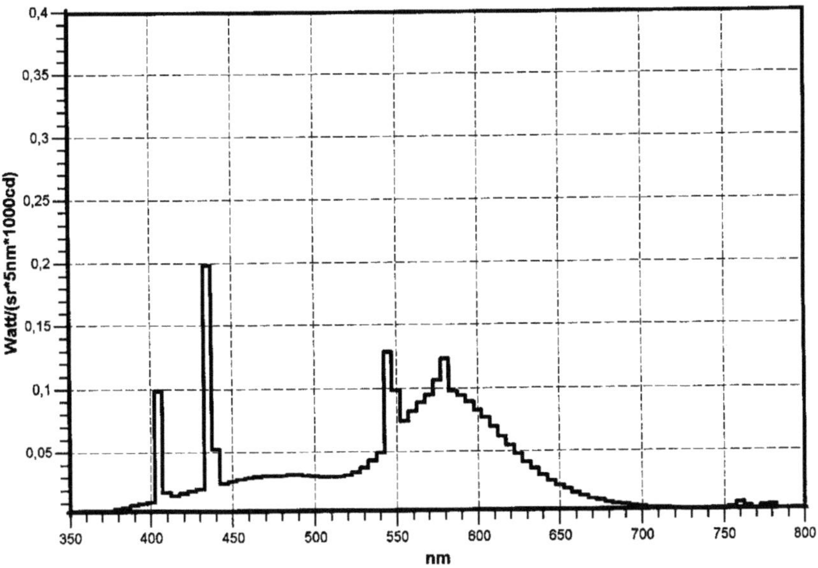

Figure 7-7. Halophosphate 4000K (color "640"), CRI = 60.

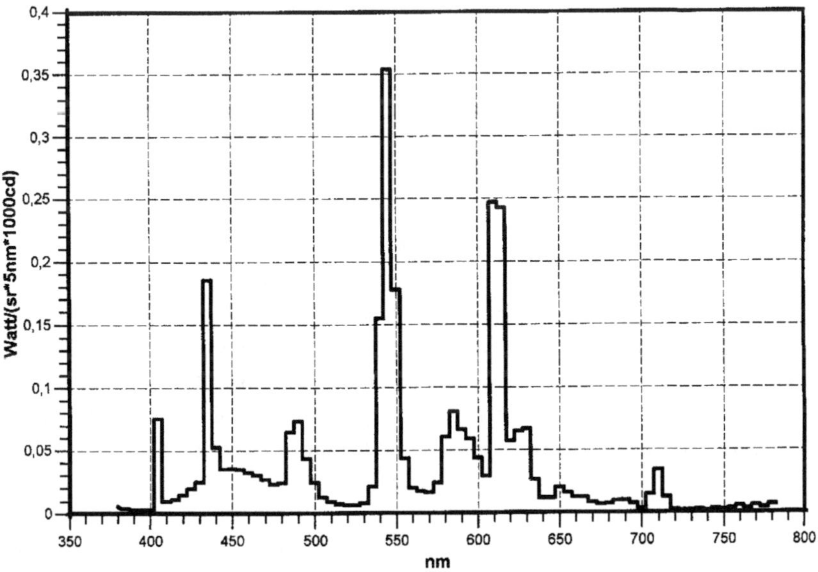

Figure 7-8. Triphosphor 4000K (color "840"), CRI = 85.

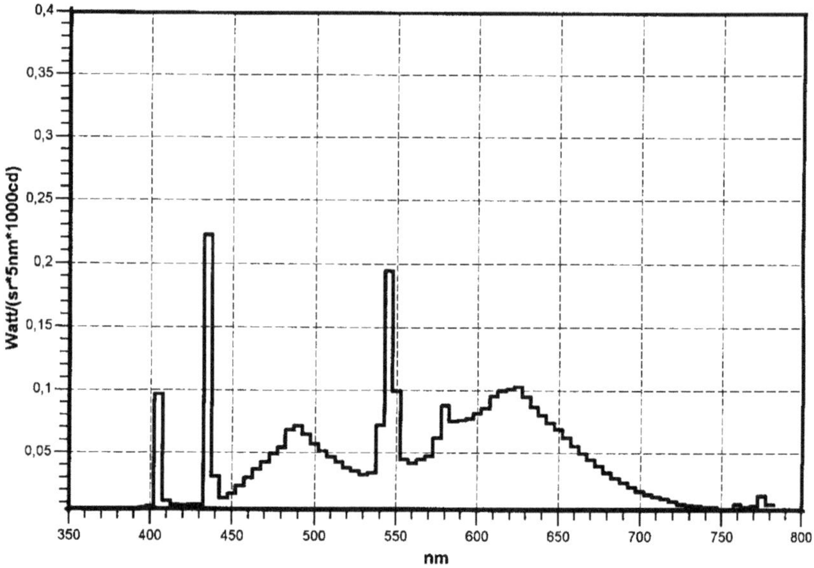

Figure 7-9. Color "940," CRI > 90.

the diameter, an optimal current had been determined that yielded a maximum efficacy. Fluorescent lamps with 5/8, 8/8, and 12/8 inches (16, 25 and 38 mm) tube diameter, called T5, T8 and T12, and of different lengths, were designed.

Within a short time, the famous 48-inch-long 40WT12 lamp with a current of 0.43A and a voltage of 100V became the most popular "workhorse" among the fluorescent lamps, in the United States as well as in Europe. The smaller-diameter lamps could not be as highly loaded as the 40W lamp because they did not operate under optimum thermal conditions. Further, these lamps produced a high lumen drop over life.

The "Energy Crisis" in the early 1970s gave reason to review the lamp data and look for possibilities of reducing the power consumption of the lamps—and, if possible, without a reduction of light output.

From that time on, the further development of fluorescent lamps followed different directions between the United States and Europe.

United States

Because of the Energy Crisis, it was necessary to look for an improved lamp that would operate at lower wattage on existing ballasts. The solution was a lamp with krypton filling gas (or a Kr mixture) instead of the argon, hitherto nearly exclusively used. It was already known that krypton reduces the lamp voltage.[30]

It is a property of the rapid start and sequence start ballasts mainly used in the United States that their output current is nearly independent of the lamp voltage. Therefore, any voltage reduction will reduce the power consumption by nearly the same percentage. The cost of those lamps—called Energysaver, Wattmiser, Econowatt, etc.—was higher than that of a standard 40W. Not so much due to the higher price for the krypton, but due to an essential transparent, electrically conducting layer inside the glass tube that was needed to assure reliable lamp starting on existing rapid start ballasts. Nevertheless, it could be easily shown that the energy cost savings to the customer over the life of the lamps were many times the extra cost of the new lamps.

Europe

In Europe, the fluorescent lamps are mainly operated on a preheat circuit, comprising a magnetic choke and a glow starter. When the lamp voltage is lowered, this circuit reacts with a distinct current increase. The consequence is not only that the lamp wattage drops by a lower percentage than the lamp voltage but, in addition, the higher current causes higher losses in the choke. It could be shown that this nullifies the wattage savings in the lamp—it even overcompensates it.

A short time later, the fill gas had been changed also in Europe from argon to a mainly krypton-containing mixture, but for quite a different reason:

In contrast to the American market, in Europe there was and is much more interest in lamps with better color rendition. Therefore, the question was how to introduce the new tricolor lamps at an acceptable price into the market despite the expensive new phosphors. At the beginning their price was up to 50 times that of halophosphate lamps!

Several investigations were started to lower the phosphor costs for the tricolor lamps: the emphasis was on getting cheaper raw materials, reducing phosphor preparation costs, optimizing the grain size distribution and the coating procedure (including splitting the single-phosphor layer into a halophosphate and a triphosphor layer) in order to reduce the weight of the expensive tricolor phosphors. Of course, any cost reduction must not reduce or degrade lamp performance!

In time, all of these measures were successful. One of the most attractive steps was to lower the phosphor weight by reducing the lamp diameter, particularly as there were some hints that this also could help to increase the efficacy of the lamp. Intensive investigations did confirm this, and showed that the efficacy really has a maximum between 25 and 30 mm tube diameter. Therefore, the target for a new lamp series was T8 (25 mm) to save phosphor and to use an existing glass tube design.

From T12 to T8, A Market Revolution

With the stable and highly loadable tricolor phosphors, it was possible to reduce the lamp diameter without getting into the danger of considerable blackening. The lumen maintenance also remained excellent.

But another consequence had to be taken into account: Decreasing lamp diameter without changing other lamp parameters increases the lamp voltage.

For the design of a new fluorescent lamp series, two concepts arose—to stay with the conventional argon fill gas, or to change the fill gas in order to compensate for the voltage increase by using the voltage-lowering property of krypton:

Argon Lamps

The higher lamp voltage would increase the power consumption, or else new ballast designs would be needed. Both were not good arguments for a market introduction. In addition, at undervoltages of the 220V line of the European network, no stable operation of the lamps could be guaranteed, especially with the European 60 in. (1500 mm) long lamps.

Krypton Lamps

In a preheat choke/starter circuit, an increase of the lamp voltage leads to a higher lamp power consumption, whereas a decrease leads to a higher power loss in the choke. Therefore, it was obvious that a T8 lamp, when replacing a T12 lamp, should have exactly the same voltage as the T12 lamp. Due to the electrical characteristics of the choke, the current (*cum grano salis*) should also not change.

An additional and important consideration was that the ballasts did not need to be reapproved if they operate lamps with the same electrical characteristics.

In Europe, these considerations led to the new T8 series with nominal lamp powers of 18W, 36W, and 58W instead of the T12 series with 20W, 40W and 65W versions and the same lengths (24, 48, and 60 in., or 600, 1200 and 1500 mm).[31,32]

An explanation is needed about why the T8 lamps consume 10% less power, although their voltage and current is the same as the respective T12 lamps. The reason is because the higher re-ignition voltage peak of the T8 lamps reduces the lamp power factor.

It should be mentioned that the power factor of a line operated fluorescent lamp is less than unity, not because of a shift between current and voltage (both pass zero at the same time!), but because of a waveform distortion. The voltage wave shape of the lamp deviates strongly from sinusoidal (see Figure 7-10). It could be shown by Fourier analysis of the wave forms, that the stronger deviation of the T8 lamp decreases the power factor of this lamp and leads to the 10% lower power consumption.[33]

There is one disadvantage of the T8 krypton lamps compared to the argon version: The lamp operates somewhat "jittery" in the first few seconds after it is switched on, especially if the lamp temperature is under 20°C. Fortunately, fluorescent lamps are mainly used for indoor lighting. If used outdoors, the effect could cause some problems, but not if electronic ballasts are used.

After intensive review of all information, it was decided to accept this disadvantage, because it is highly overcompensated by many important advantages:

- Problem-free interchangeability with T12 lamps in preheat circuits;
- 10% saving of lamp power;
- No additional ballast losses;
- As bright as a T12 lamp (the efficacy of a T8 lamp is 8% over the respective T12 lamp; in common fixtures, the T8 lamp re-absorbs less light because of its smaller diameter—if this is 2% more, the T8 fixture provides the same light output as a T12);
- Less phosphor consumption;
- Less glass consumption;
- Less energy consumption to produce the lamps;
- Less packing material;
- Lower weight; and
- Lower volume needed to transport and to store glass tubes and lamps.

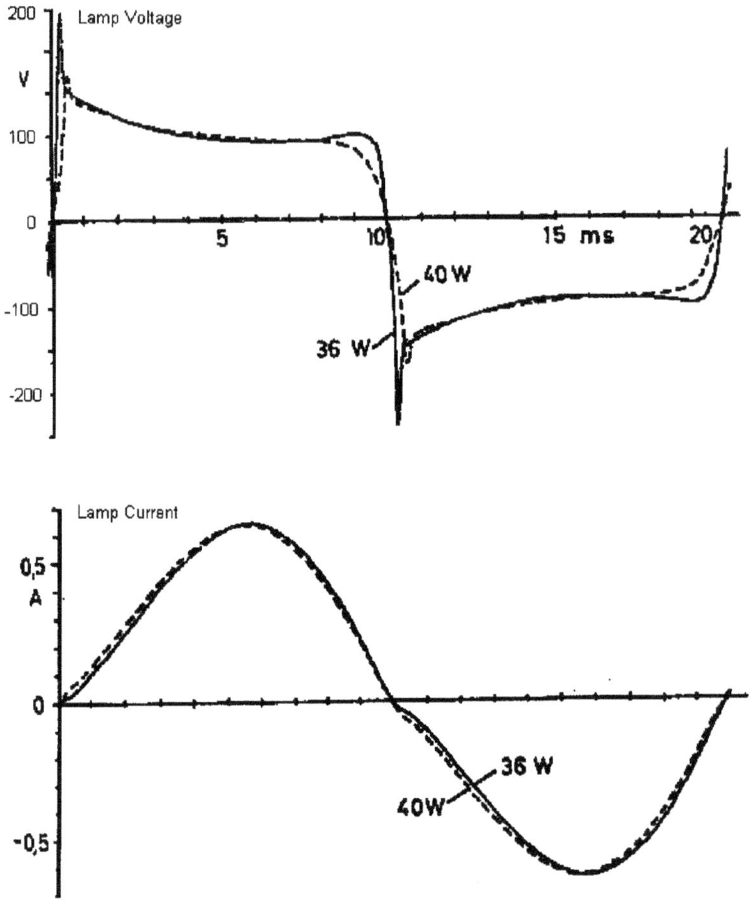

Figure 7-10. Voltage and current curves of a 40WT12 and a 36WT8 lamp.

All these reasons contributed to the new lamps finding excellent acceptance in the European market. But the most important fact was that the lamp manufacturers learned how to improve the procedures to prepare and to process the halophosphate phosphors. Thus, it was soon possible to offer the new T8 lamps not only in the triphosphate version, but also in halophosphate and other standard colors that are still the major type of all fluorescent lamps sold.

As a consequence, within about one decade, nearly the whole European fluorescent lamp market is being converted from T12 to the ecologically and economically advantageous T8 lamps.

In the American market, the T8 lamps are also on the advance. But this development is delayed because it is not possible to replace T12 lamps in America in the same easy way as in Europe. Nearly all lamps here are operated on rapid start or similar ballasts. These ballasts cannot operate the T8 lamps without changes. That means, together with the lamps, the ballasts also must be exchanged. This is an expensive procedure, and may be economical only if the ballasts and/or fixtures are near the end of their lives.

It is anticipated that the situation will change with the growing introduction of electronic ballasts.

The Principles of Electronic Ballasts

In a gas discharge lamp, a voltage pulse makes the gas conductive, usually with the help of an ignitor or starter. A magnetic inductance ballast is usually used to limit the current. Electronic ballasts operate lamps by substantially increasing the frequency. This offers a great advantage of not only reducing the size and weight of the inductance, but significantly lowering its energy losses.

A typical example is shown in Table 7.1 for the standard 36 or 40W preheat fluorescent lamp.

Table 7.1. Properties of chokes at low and high frequency.

Frequency	*50 Hz*	*50 KHz*
Impedance	390 Ω	390 Ω
Inductivity	1 henry	1 mhenry
Apparent choke power	72 VA	63 VA
Choke power factor	0.05-0.15	0.005-0.015
Losses	3-10 watts	0.3-0.9 watts
Volume	150 cm^3	15 cm^3
Weight	750 grams	25 grams

In addition, the high frequency operation of the lamps significantly improves their efficacy, as seen in Table 7.2. For the 36 or 40W preheat lamp, the gain amounts to 22 percent. Overall, the gain in this and other fluorescent lamps provides a major improvement in the efficacy of light generation.

Table 7.2 Improvement in efficacy for a standard fluorescent lamp through the use of an electronic ballast.

	50 Hz preheat circuit	*50 KHz electronic ballast*
Lamp loading	36 watts	32 watts
Light output	3400 lumens	3300 lumens
Efficacy	94 lpw	103 lpw
Ballast losses	8 watts	3 watts
Overall efficacy	77 lpw	94 lpw
Improvement in efficacy	—	22%

Even greater efficacy improvements are obtained when electronic ballasts are applied not to replace the simple choke of a preheat circuit but to replace more sophisticated devices like Rapid Start or Sequence Start circuits with their additional losses in connection with electrode heating or higher output voltage generation.

The heart of an electronic ballast is a high frequency (HF) generator. There are numerous designs of transistor circuits which have very low losses, often as little as 5%, i.e., very high efficiencies. The HF generators are widely adaptable to all the various lamp types and to special needs of lamp operation (such as low temperature starting or dimming operations).

The DC source needed to drive the HF generator may be a simple, very basic bridge rectifier and a smoothing capacitor. However, rather than being preferably sinusoidal, the line current waveform delivered by such circuits tends to be needle-shaped and contains a high percentage of harmonics that far surpass the limits established for illumination purposes (IEC Publication 555-2). Special devices are therefore incorporated into electronic ballasts to provide line currents with low distortion (PF correction). In addition, means for suppression of the high frequency disturbances on the line itself are included (HF suppressor).

To deliver all of these features, a high quality electronic ballast must be of a much higher complexity than a simple high frequency

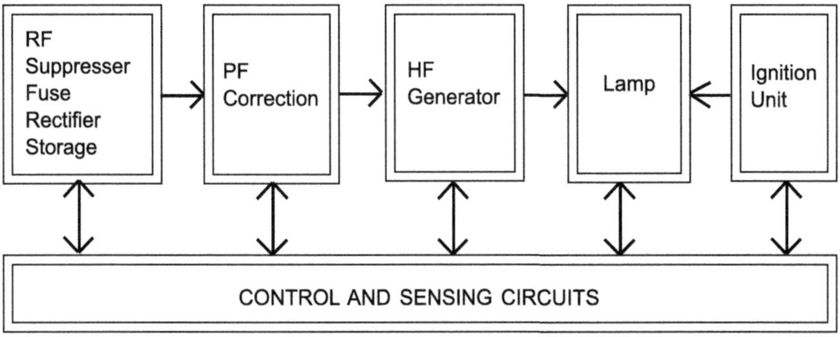

Figure 7-11. The fundamental elements of electronic ballasts.

converter. Figure 7-11 shows the diverse nature of the multiple circuits that are required.

At the onset of electronic ballast development, very little was known about how the transformation into a sinusoidal waveform could be achieved. Now several methods are available. The simplest way is the use of a magnetic low-pass filter but it is not the most practical way because of its large size and excessive filter losses. Current methods basically control the charging of the smoothing condenser. A widely used circuit, shown in Figure 7-12, is the so-called " boost converter" in which the charging is regulated by a transistor, and where the output voltage is higher than the input voltage (it is boosted).

Figure 7-13 shows how capacitor, C2, is charged via a storage inductance by producing, with the help of a switching transistor, many short pulses during each half cycle to give an integral sinusoidal shape to the current waveform.

This circuit is of great significance because the characteristics of the current waveform can be easily controlled by an IC. The integrated circuit can also be modified to include systems that can optimize preheating and dimming, either manually or with external signals.

A very simple method to lower the distortion of the line current is to omit the electrolytic smoothing capacitor, C2, and the other booster components. In this case the HF generator operates directly behind the bridge rectifier and the line current is essentially sinusoidal. However, the HF lamp operating current is deeply amplitude-modulated, and the re-ignition losses nullify the advantage of the HF driven lamp, described in the next section.

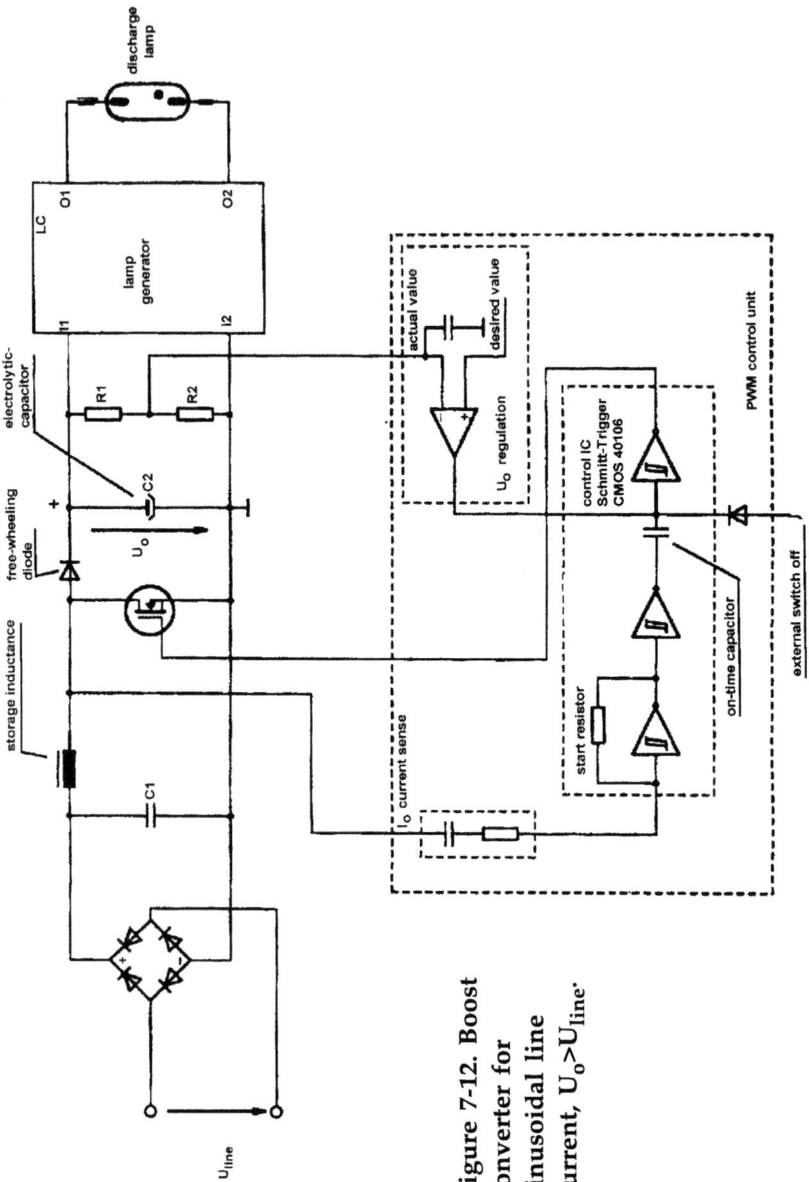

Figure 7-12. Boost converter for sinusoidal line current, $U_o > U_{line}$.

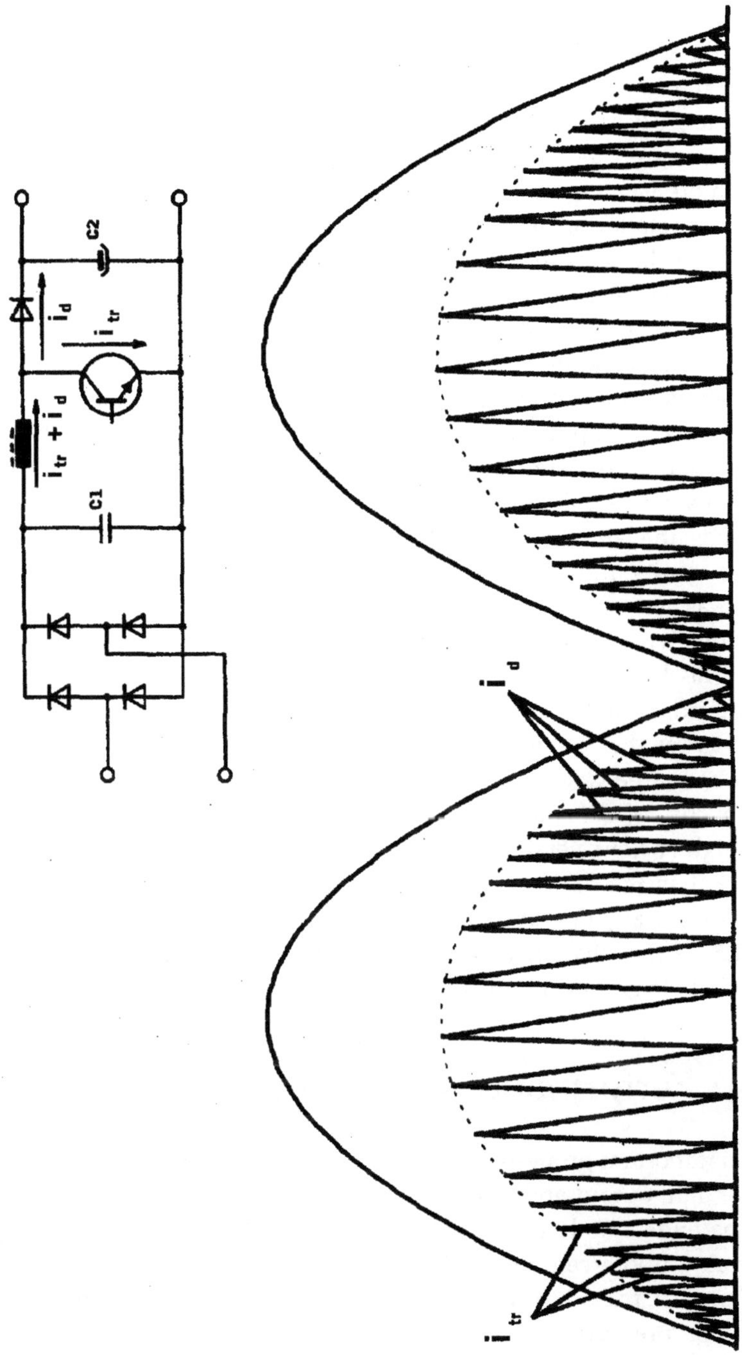

Figure 7-13. Charging of smoothing capacitor and line current.

High Frequency Operation of Fluorescent Lamps

In the 1980s, electronic ballasts for lamps came into the market, allowing the operation of fluorescent lamps at much higher than the 50/60 Hz line frequency. A special advantage of fluorescent lamps is that their efficacy rises by about 10% if the frequency is increased to some 10 kHz or more (see Figures 7-4 and 7-14).

The explanation is that at high frequencies, the AC period is shorter than the relaxation time for ionization and de-ionization. That means the ionization density in the discharge column remains nearly constant over the whole AC period, and therefore stays closer to the optimal operating conditions for which the lamp is constructed. In contrast to this behavior, the ionization density at low frequencies oscillates more or less around the optimum, causing poorer discharge conditions in the average. In addition, after each half-period of the lamp current the discharge must be re-ignited which costs energy. Due to the smaller density oscillations, the lamp behaves at high frequencies like an ohmic resistor, that also results in an increasing power factor (see Figure 7-15).

But as aforesaid, there is another advantage of a modern electronic ballast. Its own losses are lower than those of a conventional magnetic ballast. This helps to increase the efficacy of the lamp-ballast system.

A third advantage of the electronic ballasts is that they can easily operate lamps that cannot be directly operated by a choke on the line, because of a large lamp voltage. Examples of this possibility are the new T5 lamps, up to 1500 cm long. They are on the market in a low-current and a high-current version. The low-current version has even higher efficacies than the T8 lamps.

Electronically operated fluorescent lamps with even smaller diameters (T2 = 7 mm diameter, or even T1) have been designed for special applications (e.g., in cars, furniture, laptop computers).

Lamp Performance and Temperature

An important property of fluorescent lamps is their specific dependency of light output and electrical performance with temperature. This has to be understood and recognized to get the best performance possible from a fluorescent lamp.

The basic reason for this behavior is the strong change of the mercury vapor pressure with temperature. Each lamp contains a small

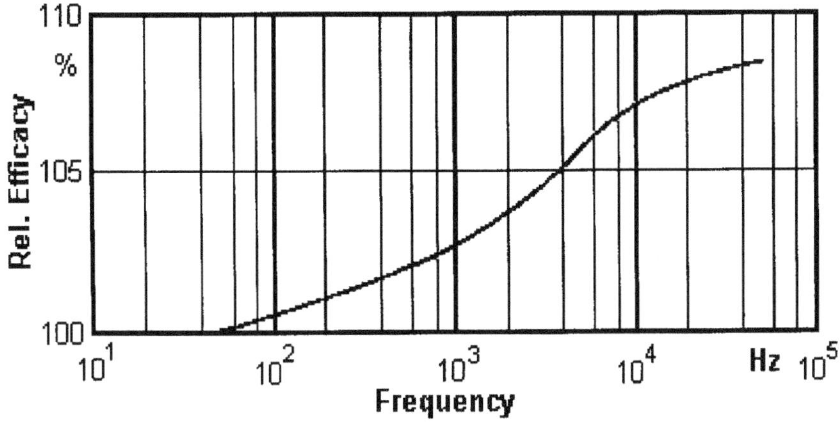

Figure 7-14. Higher frequency operation increases the efficacy of fluorescent lamps.

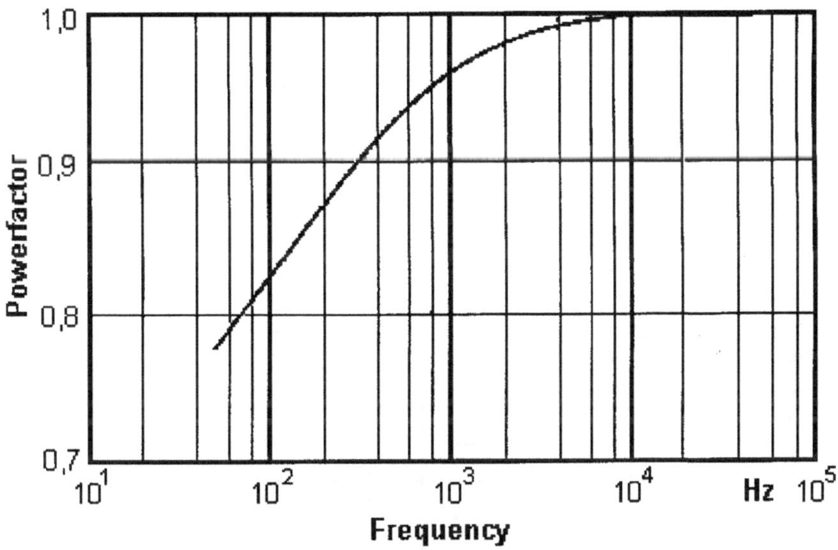

Figure 7-15. Increase of lamp power factor with increasing frequency shows the ohmic behavior of a high-frequency discharge.

amount of mercury (nowadays only a few milligrams).

Mercury is needed, because it is by far the best element to generate UV with high efficiency in a low-pressure discharge. Many investigations have been made to get the same good results without mercury, but all alternative systems were found to suffer from much lower luminous efficacies than a lamp with mercury—lower at least by a factor of two.

Nevertheless, mercury-free lamps are of interest. They are completely free of the mercury-specific strong performance relationship with temperature. This means, too, that these lamps do not vary their light output or electrical characteristics during "run-up," when the lamp temperature increases from the "cold" (non-operating) to the operating state. This, for instance, can be important for photographic or electronic reproduction devices that may need a light source with a highly constant lumen output immediately when switched on.

Typical curves of light output versus temperature of fluorescent lamps are shown in Figure 7-16. The curve on the left is valid for the "normal" case, a lamp containing just mercury.

At low temperatures, nearly all the mercury is condensed as a solid body or as liquid droplets, and the density of mercury atoms in the gas phase is very low. Therefore, the electrons have very little chance to excite mercury atoms, and consequently the luminous efficacy of the lamp is low. With rising temperature, more and more mercury evaporates. This explains the steep increase of the curve towards higher temperatures.

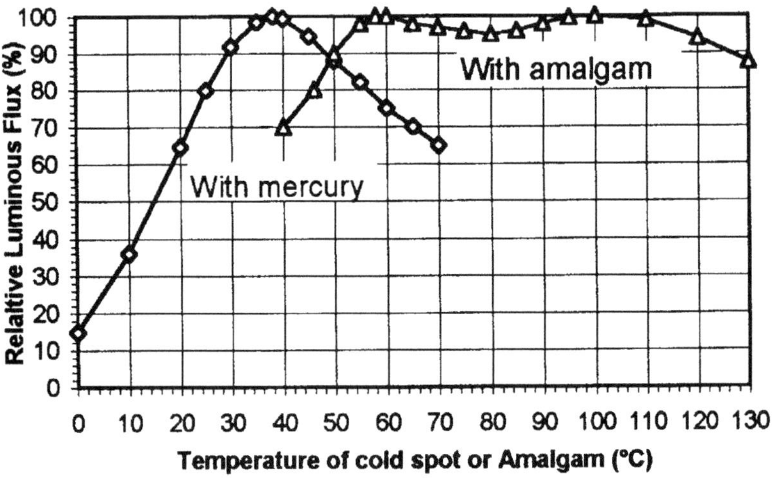

Figure 7-16. Luminous flux versus cold spot or amalgam temperature.

After reaching a maximum at about 40°C, the light output drops again. This is explained by the increasing self-absorption of UV photons. These photons can be reabsorbed if they hit other non-excited atoms that become excited in turn, and so on. Regrettably, there is a certain chance of losing the excitation energy at any impact. The energy is then transferred into kinetic form, which means heat. This leads to a reduction of photon yield.

The best condition to generate as many photons per watt as possible is provided with a temperature of about 40°C at the "cold spot"—the place in the lamp with the lowest temperature. Like dew condensing in the coldest region of a meadow, the mercury condenses at the cold spot, and its temperature controls the mercury vapor pressure in the whole lamp. At 40°C, the mercury vapor pressure is 0.8 Pa (= 0.008 millibar or 0.006 torr), the optimal value for a T12 lamp (a bit higher for lamps with smaller diameter).

Amalgam Lamps

The curve on the right in Figure 7-16 is representative of an amalgam-loaded lamp, that contains not only mercury, but an alloy of mercury with other metals, soluble in mercury. This material, called "amalgam," has a lower mercury vapor pressure than the pure mercury, or—and this is important for the application in lamps—it yields the optimal vapor pressure at any temperature above 40°C, depending on the composition of the amalgam. Not only the maximum point, but the whole lumen versus temperature curve, is shifted to the right.

If the lamp temperature always exceeds 40°C due to lamp design or applicability, the amalgam technology helps to yield the optimum vapor pressure and maximum light output. Two development targets may lead to lamp temperatures higher than 40°C. The first is the desire for an enhanced lamp loading, in order to get more light from a lamp without enlarging it. This leads to the so-called HO and VHO lamps (high output and very high output with 0.8 or 1.5A lamp current instead of 0.43A in the standard 40WT12). The other is to operate and optimize a lamp for a warmer environment—for instance, in small or hermetically closed fixtures. For both situations, amalgam technologies have been successfully developed.[34,35]

For example, by using indium amalgam it was not only possible to shift the lumen versus temperature curve to higher temperatures, but in addition the curves become wider. This means that the lumen drop is smaller, when the temperature deviates from the optimal value.[36,37] This widens the applicability of amalgam lamps.[38]

A thermodynamic investigation, based on the second law of thermodynamics, explained why indium amalgam brings about a wider lumen-temperature curve. In the temperature region in question, this amalgam separates into a solid and a liquid phase, both in thermodynamic equilibrium with the vapor phase.[39] Later on, other amalgams were found, that yielded even wider curves than indium amalgam.[40,41] The right curve in Figure 7-16 represents the result of this improved amalgam technology.

As pointed out before, the lumen output is a function of the cold spot temperature. But in an amalgam lamp, the *amalgam* controls the mercury vapor pressure. Lamp behavior is therefore determined by the temperature at the place where the amalgam is located. This need not be at the cold spot. The amalgam may have a higher temperature. However, an important precondition is that the mercury vapor pressure of the amalgam must be lower than the vapor pressure of pure mercury at the cold spot temperature. Otherwise, the mercury vaporizes from the amalgam, and travels through the discharge vessel to the cold spot, where it condenses. The consequence is that the lamp loses the properties of an amalgam lamp and operates like a standard mercury lamp.

Heat Sink Lamps

An alternative method to reduce the mercury vapor pressure in a thermally overloaded lamp is to provide the lamp with a heat sink. The lamps A, B, C and E shown in Figure 7-17 are typical examples.

The Power Groove (A) by General Electric and the Power Twist lamp (E) by Duro-Test do not use a simple cylindrical glass tube as the discharge vessel, but the tubes are deformed in characteristic manners to cause deviations from the circular cross section. Those parts of the tube farther from the lamp axis are cooler than the others, and can act as cold spots for a condensation of mercury with a temperature near the optimal 40°C, if the lamp is properly designed.

An early Philips solution for a high-loaded lamp was the Cone lamp (C), being deformed only in the middle by a cold spot in form of a small cone sticking out of the lamp. If a highly loaded lamp, operated in a horizontal position, is to be thermally adapted to room temperature, then the cone must point to the floor. If the lamp is axially turned by 180°, so that the cone points to the top, the cooling effect vanishes, and the lamp yields its maximal efficacy at a lower temperature, more useful for outdoor lighting. This was an advantage, but the lamp was not easy to handle because the cone was susceptible to breakage.

The lamp (B) in Figure 7-17 (Cold Flare lamp, Sylvania) does not show its high-loading feature. This lamp looks like a standard 40WT12 when it is not switched on. However, when lighted, the lamp ends are darker than those of a standard lamp, because the electrodes (or at least one of them) are farther away from the lamp bases. Therefore, the space behind the electrode remains cooler which has the same effect as with the other cold spot solutions.

Compact Fluorescent Lamps

Up to this point, we considered a fluorescent lamp to be a straight tube with a certain diameter and length. In many cases, however, their long length restricts their potential applications. For this reason, techniques were developed to bend the tubes, either to a "U" or to a ring. Ring-shaped lamps, "Circlines," became popular in the United States,

Figure 7-17. High-loaded lamps, vapor pressure controlled by heat sinks (A, B, C, E) and amalgam (D).

and more so in Japan. U-shaped lamps were first introduced in Germany.

In the middle of the 1970s, the oil shortage gave rise to considerations of how to save energy. One of the ideas was to replace incandescent lamps with discharge lamps and take advantage of their much higher efficacies—not only in traditional application fields such as offices, manufacturing facilities, warehouses, shops and street lighting, but also in homes, hotel rooms and all other spaces where relatively inefficient incandescent lamps had traditionally been used. Among the questions raised were: Are fluorescent lamps suitable for these purposes? How can they be reduced to the dimensions of incandescent lamps, and what could be substituted with minimal efficiency loss?

There was no lack of proposals on how to reduce the size and change the shape of fluorescent lamps, even some rather exotic ones. The ideas that survived included lamps that were made from a single or several glass tubes, being bent or folded several times. Among the first were small circular lamps, some of them with an integrated ballast in the center of the lamp, at first a magnetic one (Circlite, GE),and later with an electronic ballast (Circolux EL, Osram, 1983, Figure 7-18).

But the line of compact lamps that grew to become an extended family was not made from circular tube elements, but from small U-formed elements made either by melting two straight tubes together so that the ends form an H (Philips PL series, 1981), or bent to a U shape (Osram Dulux S series, 1982). A few years later, in 1984, a second series of compact lamps was added, with two, and then three, U-tubes instead of one. All these lamps were made from "T4" glass tubes (13 mm diameter). A third series, made from T5 tubes (16 mm diameter) followed. These lamps were not truly compact lamps to be used in place of incandescent lamps. They were developed because fixture designers asked for lamps similar to fluorescent lamps, but smaller.

All these lamps became so successful that several other lamp manufacturers adopted this technology in 1985.

The reason for this success was because the relatively high lamp loading was combined with high efficacies and compactness. Although they are operated far beyond their optimal thermal conditions, the mercury vapor pressure in each lamp type was optimized. The secret was that "cold spots" were created by special forming of the lamp ends that act as heat sinks designed to meet the optimal 40°C cold spot temperature.

To further promote energy savings, compact fluorescent lamps

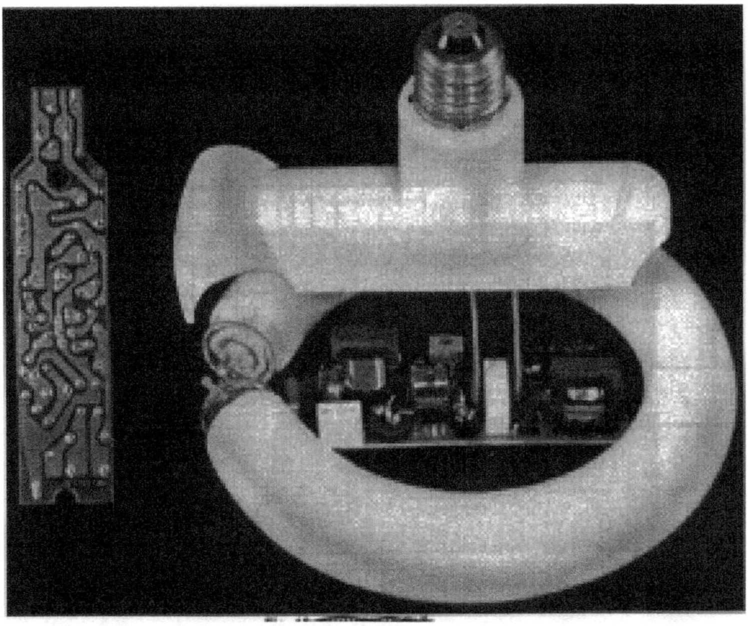

Figure 7-18. An early model (1983) of an electronic-ballasted compact fluorescent lamp. Courtesy of Osram.

were introduced that could be directly screwed into a medium base socket like an incandescent lamp. The ballast therefore needed to be integrated with the lamp, in this sense, the lamp was self-ballasted. Circlite and Circolux are examples of such lamps, but their dimensions were still too large. They could replace an incandescent lamp only in large diameter fixtures or table lamps.

The first successful self-ballasted screw-in compact fluorescent lamps were the Philips SL (since 1980) with a double-folded T4 discharge tube, a choke and a glow starter in a closed glass vial.

As the temperature inside the vial exceeds 40°C, the amalgam technology is used to control the mercury vapor pressure.

An important prerequisite for the successful SL, and all the other compact fluorescent lamps, was the availability of the highly stable tricolor phosphors.[24,25,26,27] These phosphors allowed to reduce the tube diameter down to T4 (and even smaller) without causing an unacceptable high loss of efficacy over the life of the lamp.

The first small compact fluorescent lamp with an electronic instead of a magnetic ballast was introduced in 1985 by Osram (see Figure 7-19). It took much developmental effort to overcome the temperature problems affecting the components of an electronic device located very close to the heat-producing lamp. Additionally, this effort was directed to lowering the costs for the electronic circuit so that the lamp could be sold at an acceptable price. Philips Lighting Company followed with its own design in 1987.

From a commercial point of view, it is assumed that customers will use the self-ballasted lamps like an incandescent lamp and will expect their life to be nearly independent of the number of times the lamp is switched on and off. Therefore, it was essential to develop a ballast that starts the lamp so smoothly, that nearly no emission material is sputtered or evaporated from the electrodes. This work has resulted in lamps the life of which is nearly independent of the ON/OFF switching cycle. Typically, these lamps can withstand many hundred thousands of

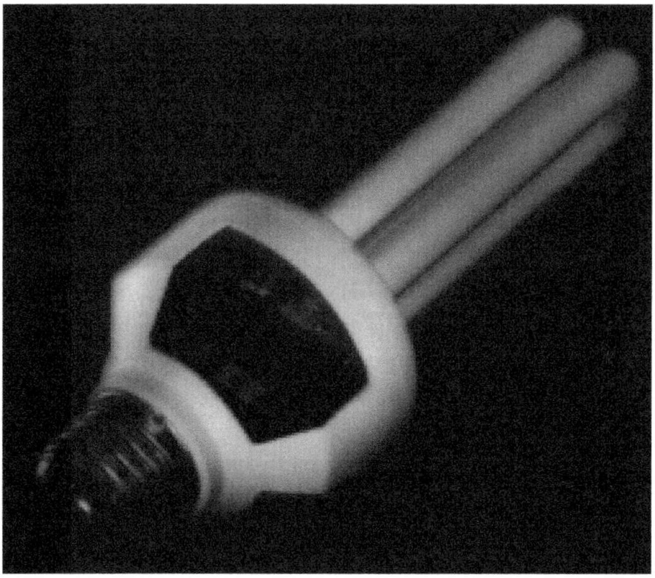

Figure 7-19. The Delux EL lamp. The first compact fluorescent lamp to use an electronic ballast and designed as a replacement for incandescent lamps. Since then, the popularity and prolific use of this and similar lamps has helped to reduce energy consumption throughout the world.

switchings without loss of life. This proves that the often-reported strong sensitivity of the fluorescent lamp to switching is not necessarily typical for this lamp type. The problem has been avoided by improving the design of the ballast. Performance differences between high quality/ high cost lamps and low priced self ballasted lamps on the market are reflected in their switching behavior or in lamp life.

The Circuitry of Compact Fluorescent Lamps

The configuration of electronic ballasts for compact fluorescent lamps is similar in principle to those for standard fluorescent lamps. However, due to the low loading, the requirements for limiting line current distortions and for high frequency suppression are much less stringent. The circuit is fundamentally that of a high frequency generator, usually constructed of two switching transistors S1 and S2, an inductance L and a capacitor C, as shown in Figure 7-20.

The high frequency is produced by switching S1 and S2, alternately on and off. Opening S1 while closing S2 lets a current pulse flow through the lamp, thus charging the L and C components. By closing S1 and opening S2, L and C discharge through the lamp, producing another pulse. The control of the two transistors can by done either by self-excitation or by external oscillators. Circuits of this kind have very low switching losses. This is very important because it keeps the temperature low in the small ballast.

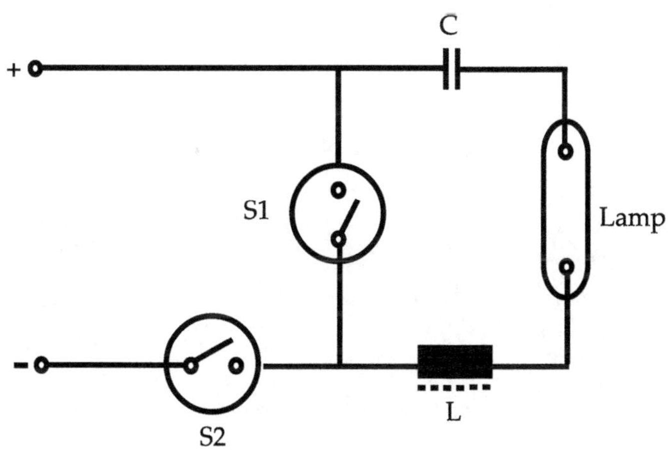

Figure 7-20. Principle of the high frequency generator used in compact fluorescent lamps.

Environmental Concerns

During his speech on the environment on the Earth Day ceremony on April 21, 1994, U.S. President Bill Clinton showed a compact fluorescent lamp as an example of saving energy—and additionally of saving jobs through the production of such energy-efficient lamps (see Figure 7-21). Not everybody might share the opinion that this lamp is really environmentally friendly, when they hear that the energy saved results from using lamps that contain toxic mercury. Let's look inside such a lamp and compare it with other possibilities of generating light efficiently.

Principally, we do not need to differentiate between large linear fluorescent lamps and small compact fluorescent lamps. Both types contain mercury, and it has been explained that all other alternatives of producing light without mercury results in lower efficacies, typically reduced by a factor of at least two. We shall illustrate the environmental

Figure 7-21. President Bill Clinton showing a compact fluorescent lamp as an example of energy-saving technology that can save money as well as reduce pollution.

aspects using as an example a 23W compact fluorescent lamp. This lamp has a rated lamp life of 10,000 hours, and therefore consumes 230 kWh during its life. This lamp contains about 5 mg mercury. If we were to replace this lamp by a (hypothetical) "mercury free" lamp, we must (at least) double the energy consumption for the same light. This means we have to pay for an additional 230 kWh of energy.

There is, however, another aspect that must be taken into account besides the costs for additional energy. Together with the fossil fuels that are extracted from the earth for the generation of electricity, we are emitting into the atmosphere toxic substances, including mercury. Investigations conducted in Germany have shown, that, to produce 1 kWh, 0.037 mg mercury is extracted from the earth and blown into the atmosphere (The Federal Office for Environmental Affairs, *Bundesumweltschutzamt* in Berlin).[42]

Of course, this figure, valid for the typical German energy mix, will vary from country to country, and even from place to place. But let's take the 0.037 mg mercury per kWh as a calculation basis. Then it is very simple to show, that—under these circumstances—by switching over from a normal compact lamp to a mercury-free lamp, we would need additional fuel for 230 kWh and therefore, emit another 230 x 0.037 = 8.5 mg of mercury into the environment. This is more than we dose into the lamp. That means if a mercury-free lamp operates at only half or less of the efficacy of a mercury containing lamp, it would consume not only much more energy to produce the same light as a mercury containing lamp, it would, in addition, pollute the environment to a greater extent with mercury than the amount contained in the lamp.

Another and rather important example is to compare an incandescent lamp for general lighting with a compact fluorescent lamp. A 23W compact fluorescent lamp produces 1500 lumens, as much light as from a 100W bulb. If the 100W bulb is replaced by a 23W fluorescent lamp, we would save 77W x 10,000 hours = 770 kWh. Following the above calculation, we would reduce the contamination of the environment by 770 x 0,037 mg = 28.5 mg mercury, more than five times the amount dosed into the lamp.

These figures clearly demonstrate the advantage that accrues from the use of highly efficient mercury-containing lamps instead of other options.

However, it cannot be denied, that all fluorescent lamps contain mercury. Is there a solution which would help to solve this problem?

Two ways have suggested themselves and have already been successfully tried:

1. Further reduce the amount of mercury in the lamp.
2. Recycle lamp materials at the end of lamp life.

Mercury Reduction

In the early years of the fluorescent lamps, the toxic properties of mercury were not well known, or not considered as much as today. It was easier to dose large mercury drops into a lamp, up to several 100 mgs were typical.

(A special solution, observed by the authors, to provide a fluorescent lamp safely with mercury under the difficult conditions of DC operation, has been used in the trains of the DC-driven Berliner S-Bahn in the 1950s: Generally, under these conditions, the polarity of the lamps must be changed every few hours. Otherwise, all the mercury disappears by electrophoresis at one end of the lamp, leaving all other parts mercury-free and dark. The large mercury drop may be 1 gram or more. Each time, when the train started, the mercury rolled backward, and forward when it stopped. By this measure, the lamps stayed bright without a change of polarity. Hopefully, the mercury was recycled at the end of lamp life.)

To reduce the amount of mercury, it was technically necessary to learn how to separate a smaller and smaller quantity of this metal (liquid at room temperature) from a supply container and to transfer it safely without losses into the lamp. This was ultimately achieved by several different technologies. One, the traditional needle and especially disk dosers were improved, and two, the specified amount of mercury was pre-dosed into a glass or metal capsule.[43] Finally, to avoid handling the liquid mercury, it was converted into a solid compound[44,45] or embedded into a porous matrix such as a sintered iron pellet.[46]

This raises the question, what is the real minimum amount of mercury needed? Theoretically, only a few tenths of a milligram of mercury are in the vapor phase during lamp operation. However, there are several mechanisms active in the lamp which lead to a remarkable consumption of mercury during lamp life. It is for this reason that the lamp must be filled with a greater amount of mercury. Now the challenge arises how to deactivate these mercury consuming processes. There are several patents which have shown how to solve this problem. Most of them are based on auxiliary layers on or under the phosphor coating, in order to protect the glass and the phosphor from mercury attack, which is one of the major loss mechanisms.

Further Aspects

Besides mercury, the modern fluorescent lamp does not contain other "dangerous" materials. Toxic phosphors like zincberylliumsilicate and arsenates have been replaced over the years. The composition of halophosphate is that of dental enamel. A small amount of cadmium oxide, used in former times to stabilize the luminescence properties, is no longer needed. The recently introduced rare earth compounds are not known to cause any harm. Lead, as far as it is still in the tubing glass, is so tightly bound that it is no real danger for the environment. Lead-containing solders have been replaced with contact pins that are welded or crimped.

Much success has been achieved in the last decades in improving the processes for the production of lamp material and parts and the lamps themselves. The method of phosphor coating of the glass tubes has been changed. The original use of organic solvents has been converted to water instead. Molybdenum, used as mandrel wire for the electrode coils has been replaced by iron, and where this is not possible, the molybdenum is recycled. The sewage systems are, therefore, free of pollutants from fluorescent lamps.

Lamp Recycling

There are two aspects for recycling of fluorescent lamps at end of life. First, the mercury content should be prevented from going into the solid waste and from there into the landfill. Additionally, materials suitable for re-use should be brought back into the production process. The glass plant can re-melt certain amount of broken glass, thereby saving raw material and improving the glass quality at the same time. But before feeding the tubing glass into the melting tank, the other lamp parts—e.g., the electrode-carrying lamp ends and the phosphor—must be separated from the tubing.

Experience has shown that the phosphor can be easily blown out of the glass tube by compressed air. Most of the mercury is simultaneously blown out. The phosphor, an amount of only about 2% of the total lamp weight, should be disposed as special waste, as it is in many cases not economical to separate it from the mercury. Some mercury remains on the glass and will be vaporized in the glass tank. Methods have been developed to absorb the mercury in the chimney of the glass plant.

The Revolution Goes On

We have seen, in the first parts of this chapter, the most important properties of fluorescent lamps, luminous efficacy and lamp life, have been tremendously improved since the beginnings of the development of this lamp family. The most important revolutionary steps have been the invention of the halophosphates and the tricolor principle, the introduction of triple coils, and the electronic ballast. What could be the next steps?

What can be done to further enhance lamp efficacy? The different energy consuming mechanisms in the production of light from electricity are schematically shown in Figure 7-22. As an example, a "modern" triphosphate 36WT8 lamp operated on a highly efficient electronic ballast with only 10% loss is used. This means the ballast converts only 4 watts of power into heat, and it will take a lot of work to further reduce this loss by another one or two watts.

The remaining 36 watts must be transferred into the lamp via elec-

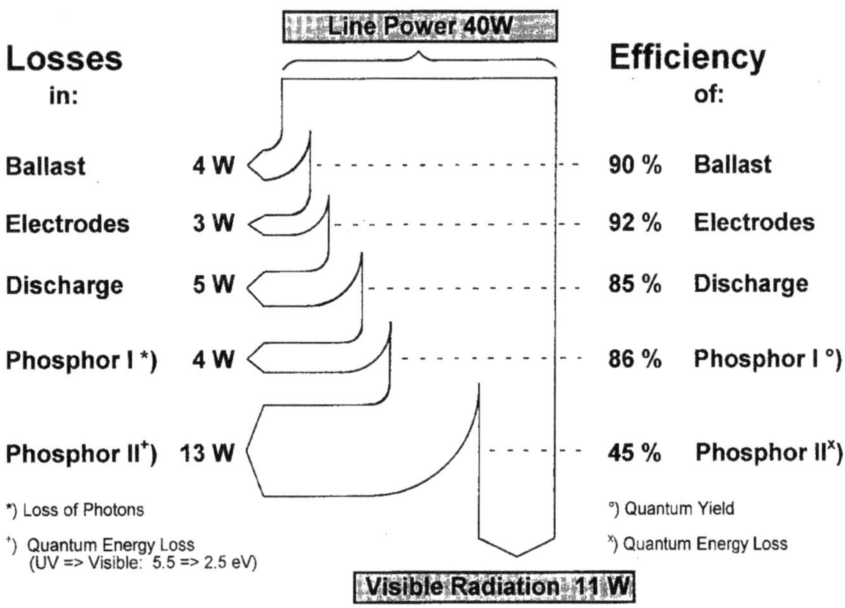

Figure 7-22. Energy flow of an electronically operated triphosphor T8 lamp.

trodes. Well activated electrodes may do this with about 92% efficiency, a further loss of 3 watts.

It was explained that a discharge with mercury produces UV photons with high efficiency (approximately 85%). This mechanism represents a further loss of 5 watts.

The last process, the conversion of UV into light, takes place in the phosphors. A good phosphor has a high quantum yield, 86%. That means 100 UV quanta absorbed generate 86 "visible" photons. This process here is designated as "Phosphor I." However, the second line, "Phosphor II," should be noted. It demonstrates that the largest energy loss (13 watts) is caused by the fact that the mercury UV lines have an average quantum energy of about 5.5 eV. Therefore, an energy efficiency of only 45% can be achieved, even if we assume that each UV photon converts into a visible photon with 2.5 eV mean quantum energy.

The last step has obviously the greatest potential for further energy savings, and many investigations had the objective of finding a system with a smaller quantum energy loss—i.e., a UV source of longer wavelength (365-400 nm) and phosphors highly sensitive in this spectral region. But up to now the great revolution has not arrived. The alternative way, to find so-called "two-photon" or "multi-photon" phosphors suitable in fluorescent lamps has not been successful. Therefore, a dramatic increase in luminous efficacy should not be expected in the near future.

For the other objective, to increase lamp life, the future has begun. Surprisingly, the basics were investigated 30 years ago. The way is obvious: In all standard and compact fluorescent lamps the life is limited by the electrode systems, as stated above. This limitation falls away for a fluorescent lamp without electrodes.

There are two known ways to couple energy to a discharge in a lamp without electrodes—by a high frequency electric or magnetic field. Philips Lighting introduced an "inductive" electrodeless lamp in 1992, based on long-known principles.[47,48] The new version of this lamp is now technically highly optimized (see Figure 7-23). GE introduced a similar lamp in 1994. Osram and Sylvania introduced another design in 1996 with the intention of applying it especially to lamps with higher lumen output (see Figure 7-24). The commercial success of these lamp types depends very much on improving the ferrite materials to reduce their magnetic losses and, in addition, the price at which the lamps can be marketed.

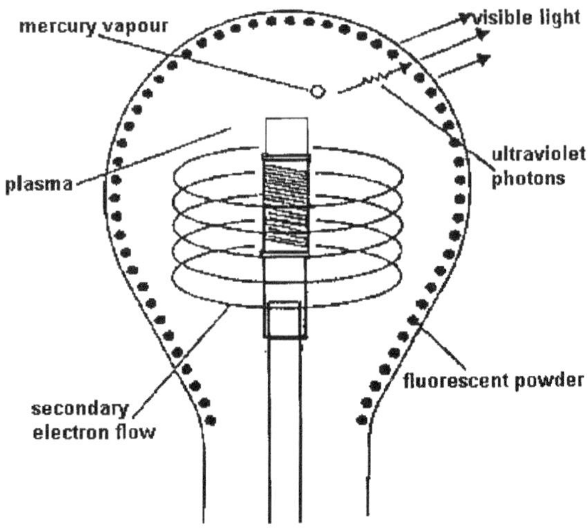

Figure 7-23. Electrodeless compact fluorescent lamp from Philips.

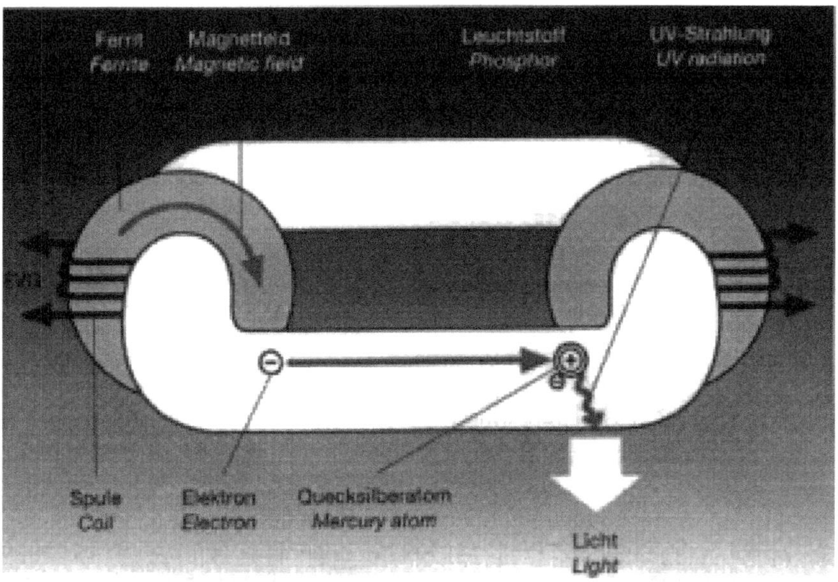

Figure 7-24. Electrodeless fluorescent lamp from Osram.

References

1 Edm. Becquerel: La Lumiere, ses couses et ses effects, Paris 1867 and 1868
2 T.A. Edison: 19.05.1896 US Pat. 865367 (10.09.1907)
3 Jacque Risler, Bull. Scient.Etud. Paris 12(1923)23
4 G. Claude: Bull.Scient.Etud. Paris 12(1923)8
5 M. Wendt: Osram-Nachrichten 32(1955)38
6 A. Ruttenauer; Licht 11(1941)4
7 Osram Nachrichten 19(1937)227
8 A. Ruttenauer, reported 26.11.1936, published in Licht 7(1937)1
9 K. Bley: Tg.d.Ges.f.Techn.Physik, Salzburg 1936
10 A. Ruttenauer: OSRAM-Nachrichten 19(1937)61
11 J.T.Randall: Light and Lighting 30(1937)
12 H.B. Ruff: Light and Lighting 30(1937)72
13 A. Claude: CIE-Conf. Paris 1937
14 G.E. Inman, R.N. Thayer: Electr.Engg.20.06.38
15 G.E. Inman: Ill.Engg. 31.08.1938
16 A.A. Bright: The electric lamp industry, N.Y. 1949
17 McKeag, A.H. and Ranby, P.W. (17.06.1942) Br. Patents 578272 and 578192, (1949) US Patent 2,448,733
18 Bodmann, Lichttechnik 11(1959)330
19 Munch, Lichttechnik 15(1959)269
20 W.A. Thornton, J.Opt.Soc.Am. 61(1971)1155
21 M. Koedam, J.J.Opstelten, Light.Res.Technol. 3(1971)205
22 K.Mahr, OSRAM Internal Report (1971).
23 K.Mahr, Electrochem.Soc.Ext.Abstr. 72-1(1972)202
24 H. Lange, British Patent GB 1004260(Nov.21, 1968), filed Feb.08,1962
25 H. Lange, R. Skibbe, Techn.-Wissen.Abh.d. OSRAM-Ges. 9(1967)157
26 J. Verstegen, J.L. Sommerdijk, J.G. Verriet, J.Luminescence 6(1973)425
27 J. Verstegen, J.Electrochem.Soc. 121(1974)1623
28 J. Verstegen, D. Radielovic, L.E. Vrenken, J.Electrochem.Soc. 121(1974)1627
29 M. Pirani, A. Ruttenauer, Licht 5(1935)93
30 G. Meister, T.H. Heine, Illum.Engrg. 44(1949)681, 47(1952)159 and 660
31 D. Hofmann, F. Schipp, Licht & Bel. 25(1978)5

32 L.E.Vrenken, Light.Res.Techn. 10(1978)161
33 D. Hofmann, Gemeinschaftstagung der Lichttechnischen Gesellschaften, Berlin, LiTG (1980)Vol.1,94
34 A. Lompe, H. Dziergwa, Patent DE 1086804, Apr. 29, 1958 ; Techn.Wiss.Abh.d.OSRAM Ges. 8(1963)78
35 K. Eckhardt, Techn.Wiss.Abh.d.OSRAM Ges. 9(1967)51
36 C.J. Bernier, J.C. Heffernan, Ill.Engineering 59(1964)801
37 G.S. Evans, P.E. Snyder, Ill.Engineering 61(1966)51
38 B. Kuhl, K. Eckhardt, Techn.Wiss.Abh.d.OSRAM Ges. 11(1973)120
39 D. Hofmann, E.G. Rasch, Techn.Wissensch. Abh.d OSRAM Ges. 11(1973)106
40 J, Bloem, A. Bouwknegt, G.A. Wesselink, J.IES (1977)141
41 J. Bloem, A. Bowknegt, G.A: Wesselink, Philips techn. Rdsch. 38,1(1979)12
42 Ralph Ebersperger, Wolfgang Mauch, HLH, 43,3 (1992)113
43 Ridders, v.d.Wolff, Mollet: Patent DE 20306 (June 27, 1969) (VDI-Verlag, Germany)
44 P. della Porta, E. Rabusin IV. Int.Symp. Resid. Gases in Electron Tubes, Fiorence (1971)
45 M. Borghi, et al.: Light.res.Techn. 12(1980)73
46 W. Schuster, Techn.Wissen.Abh.d.OSRAM Ges.12(1986)206
47 John M. Anderson, U.S. Patent 3,500,118 (March 10,1970), filed July 17, 1967
48 John M. Anderson, U.S. Patent 3,521,120 (July 21, 1970), filed March 20, 1968

Chapter 8

Mercury and Metal Halide Lamps[1]

by Alexander Dobrusskin and Jürgen Heider

Disadvantages of the incandescent lamp are its low efficacy of only about 15 LPW and the relatively short life of approximately 1,000 hours, due to the high IR radiation loss and the evaporation of the tungsten filament. During the revolution of lamp development in the '30s an innovative step to avoid these disadvantages was the development of lamps with electrical discharges in gases. The first types were HID mercury lamps, with five times the efficacy and life of incandescent lamps. These lamps were used for general lighting purposes on supply voltages of 200-240 volts. The hard glass arc tubes, later replaced with quartz, contained a small amount of mercury, which evaporates completely during the warm-up period of the lamp, developing a stable arc between two tungsten electrodes. But the mercury radiation consists of only a few strong lines in the yellow, green and blue, and very little red. To improve the poor color rendition, fluorescent coatings were used inside the outer bulbs making use of the UV radiation of the mercury spectrum of the arc.

The next significant development step, which occurred in the '60s, was the application of metal halide additives to the mercury. These compounds evaporate and dissociate at the high temperature of the arc and emit an additional multi-line spectrum. Improvements included efficacies of up to 100 LPW, good color rendition and long life. In this chapter, we will discuss mercury and metal halide lamps.[1]

Introduction

The revolution in lamp development can be shown best by looking at the tremendous increase in efficacy over the years (see Figure 8-1). The highest efficacy has been achieved with low pressure sodium (LPS) lamps. Because of the monochromatic yellow light, LPS lamps are only used in applications where it is not necessary to distinguish between different colors of the illuminated objects. Therefore, these lamps were used only for highway and parking lot lighting, mostly in northern Europe. This chapter covers only lamps with "almost white colors" and good color rendering capabilities—i.e., lamps with radiation throughout the visible spectrum.

High intensity discharge (HID) lamps are characterized by high luminous efficacies and long life, high brightness and high power concentration. The most important lamps are mercury lamps, rare or noble gas, high pressure sodium (HPS) and metal halide (MH) lamps. Currently, the HPS and MH lamps offer the highest innovation potential for the development of more economical lamps with high luminous efficacies, long life and good color rendering characteristics.

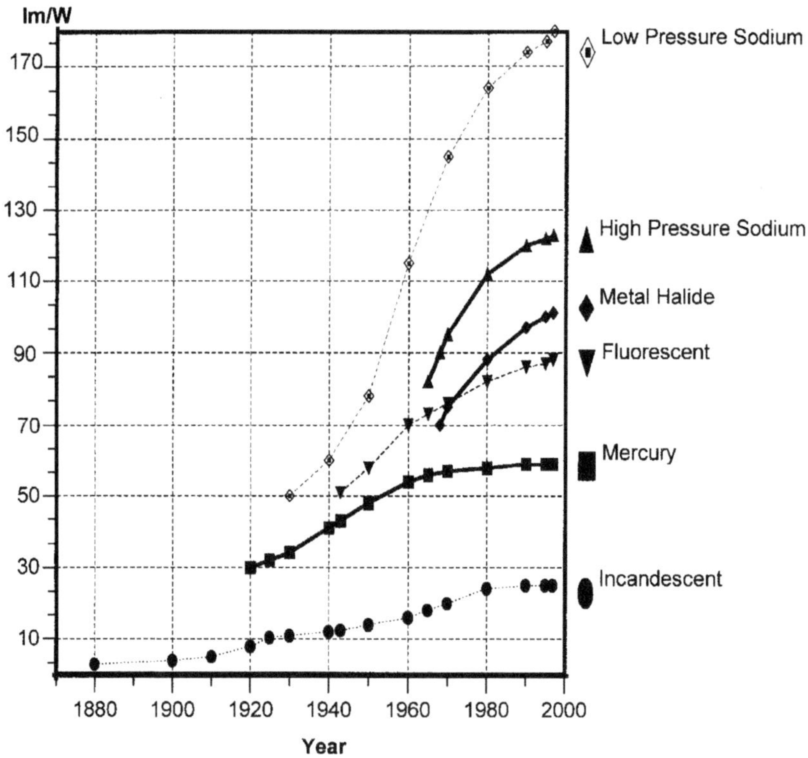

Figure 8-1. Development of the efficacy of various light sources.

The concepts and physical principles underlying HPS and MH lamps, which are grounded in discharge physics, have been known for some time. They held the promise of significantly higher luminous efficacies and vastly improved source color in comparison with high pressure mercury lamps. Scientists and lamp designers, however, realized how difficult it would be to overcome some major problems which confronted them from the materials point of view. It was necessary to prepare halides of high purity and to develop procedures to dose those very hygroscopic compounds into the discharge vessel. A tubular arc tube of quartz had to be produced from the highest purity quartz glass and equipped with high quality tungsten electrodes. These material requirements were finally met during the '60s, when the processes for very pure quartz tubing were introduced and high grade metal halides could be prepared, dosed and handled.

Principles of the High Pressure Discharge

Definition, Thermal Equilibrium

High pressure discharges are determined by their plasma equilibrium. That means the components of the plasma: neutral particles, electrons and ions have approximately the same MAXWELL speed distribution and thereby the same thermal distribution:

$$\frac{M * \overset{\pm}{v}^2}{2} = \frac{Mi * \overset{\pm}{v}_i^2}{2} = \frac{Me * \overset{\pm}{v}_e^2}{2} = \frac{3}{2}kT$$

M, M_i, M_e are the masses of the neutral particles, the ions and the electrons, $\overset{\pm}{v}$, $\overset{\pm}{v}_i$, $\overset{\pm}{v}_e$ their average velocities, k, the Boltzmann constant and T, the plasma temperature.

In high pressure arcs, the equilibrium condition is only fulfilled in small areas and during short periods of time, because there is a strong thermal gradient in direction to the wall of the discharge tube and because of the AC current, the predominant mode of lamp operation, the temperature alternates as well. Moreover, electrical and magnetic fields influence the equilibrium.

Within sufficiently small areas and in presence of thermal ionization, the thermodynamic description of the plasma has been given by the SAHA equation:

$$\frac{n_e \Sigma n_i}{n} = 2\frac{Z_i}{Z} \Sigma\left(2 \pi k T\right)^{3/2} \Sigma h^{\pm 3} \Sigma \exp\left(\pm \frac{eV_i}{kT}\right)$$

Where n_e, n_i, n represent the number of electrons, ions and neutral particles per cm^3, Z_i and Z are the partial function of the ions and neutral particles and eV_i is the ionization energy of the atoms.

The partial function is defined by:

$$Z = \sum_{a=1}^{\infty} g_a \Sigma \exp\left(\pm \frac{eV_a}{kT}\right)$$

where eV_a is the excitation energy and g_a the statistical power of the status g_a, and where

$$g_a = 2J_a + 1$$

and J_a represents the entire spin momentum of the *a* state.

In addition to the SAHA equation, other physical laws are known, which in principle enable one to compute the processes which occur in a high pressure arc. An exact description is, however, not possible because of the mathematical difficulties arising from the unknown temperature dependence of the coefficients, difficulties in defining boundary conditions and in the case of bent arcs, with many other components. In metal halide discharges, the sum of states, especially those of the molecules, are not known.

Through extensive simplifications and analogies it is, however, possible to describe simple arcs—e.g. the mercury arc.[2,3,4,5]

Processes in the Plasma

The processes important for the generation of light and for the stationary operation, the electric discharge in the plasma is governed by electrons: ionization, recombination, excitation, emission, double and triple collisions.

In metal halide lamps containing molecule vapors additional thermodynamic reactions occur—e.g., chemical dissociation and recombination of the metal halides, which participate significantly in the generation of light.

Stationary Arcs

Electrical arcs should in practice operate in a stabilized state without flicker. Usually the stabilization mechanisms are described as shown in Figure 8-2.

Wall-Stabilized Arcs

For example, consider the low pressure discharge, fluorescent lamp. The arc is diffuse and completely fills the discharge tube.

Convection-Dominated Arcs

Most of the high pressure arcs—e.g., high pressure mercury lamps or metal halide lamps. The arc is constricted and bent due to the heat convection.

Electrode-Stabilized Arcs

For example, xenon and high pressure mercury short arc lamps. The arc is concentrated strongly and operates in the short gap of the electrodes.

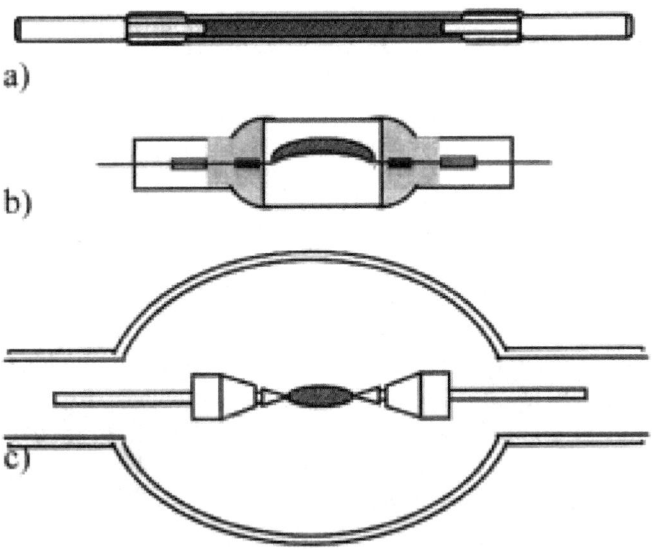

Figure 8-2. Arc stabilization in the case of high intensity discharge lamps: a) wall-stabilized discharge in water-cooled rare gas lamps; b) convection-dominated discharge metal halide lamps; c) electrode-stabilized discharge in Xenon and high pressure mercury lamps.

This classification is, however, not exact because all three stabilization effects occur in all arcs.

Arc Characteristic

As a rule, the electrical resistance characteristic of the arc is negative. For stationary operation, current-limiting means have to be used to avoid the destruction of the discharge tubes through excess heating caused by the rising current.

Regulating elements that can be used include:

- Ohmic resistances—e.g., incandescent filaments in the case of self ballasted lamps;
- Inductive resistances—chokes, lead peak transformers;
- Capacitors, usually in combination with chokes or transformers; and
- Electronic ballasts, rectifier circuits.[6,7,8,9]

High Pressure Mercury Lamps

High Pressure Mercury Lamps with Coated Bulbs

A glass outer jacket is used to protect the feed through materials of the arc tube from getting oxidized and to eliminate the harmful UV radiation. This usually ellipsoidal shaped outer bulb, which is often coated with a fluorescent powder, contains a transparent quartz glass discharge tube filled with mercury (Hg) and argon. The design of the arc tube is shown in Figure 8-3. The electrical feed-through consists of thin molybdenum foils that are pinch sealed at approximately 2200°C into the transparent quartz glass. The electrodes consist of tungsten and contain an emission material—e.g. Ba-Y-tungstate that reduces the work function of the electrodes and reduces the break-down voltage. To facilitate the ignition, the discharge tube contains, in addition to the two main electrodes, a starting electrode (probe) of molybdenum or tungsten. This starting electrode is placed in proximity to one of the main electrodes and is electrically connected to one end of a temperature-stable resistor of approximately 15-20KΩ. The other end of the resistor is connected to the lead wire of the opposite main electrode.

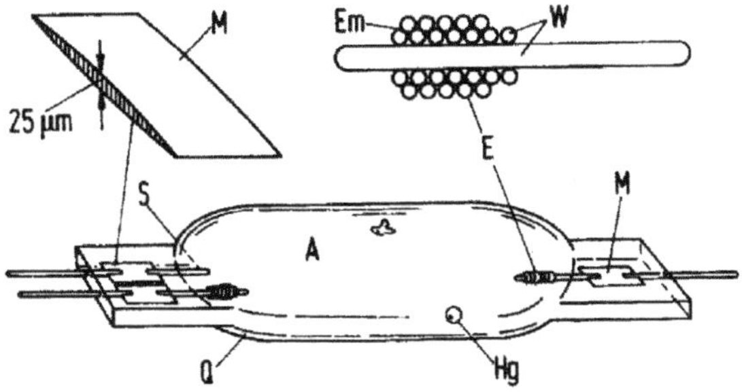

Figure 8-3. High pressure mercury lamp discharge tube: Q transparent quartz, M molybdenum foils, W tungsten, E electrodes, Em emission material, Hg mercury filling, A argon, S starting probe.

Upon applying the supply voltage, a glow discharge is established between the starting electrode and the neighboring main electrode which, in turn, activates the main discharge in the argon gas (approximately 30 mbar). The discharge tube is warmed-up within 3-5 minutes. Then the mercury is completely evaporated and the characteristic spectrum of the Hg discharge emitted. The design and function of the lamp is shown in

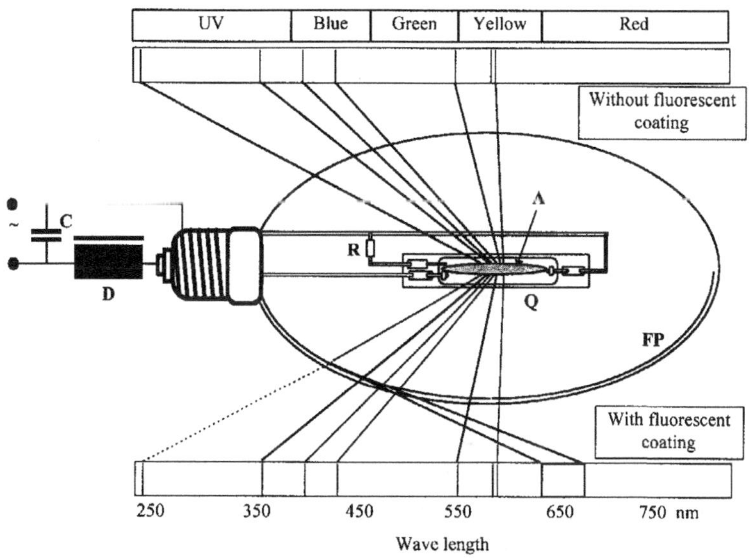

Figure 8-4. Schematic of high pressure mercury lamps: upper half without fluorescent coating, lower half with fluorescent coating with corresponding spectra. A discharge arc, FP fluorescent coating, Q transparent quartz discharge tube, D choke, R start resistor, C capacitor for power factor correction.

Figure 8-4.

The spectrum of the mercury discharge consists of only a few spectral lines. Red radiation is almost completely absent. The relatively strong ultraviolet radiation at 254 and 365 nm is therefore utilized for the excitation of a red-emitting fluorescent material—e.g., europium-doped yttriumvanadate coated onto the inner wall of the outer bulb. With this coating, the color rendering of the high pressure mercury discharge of CRI = 20 is increased to approximately 50, depending on the kind of fluorescent material and the thickness of the coating.

By using newly developed fluorescent materials, an increase in the luminous efficacy by 13-15% is possible. For comparison, the data of the 125W types are specified with three different luminous colors in Table 8-1. The warm color temperature (3000 K) and the good color rendering properties (CRI 60) of the third lamp type are attained by coating the outer bulb with an additional yellow filter that reduces the blue radiation. The better color properties are bought, however, at the expense of a lower efficacy.

Table 8-1. High pressure mercury lamps: Type HQL 125W with various color temperatures.

Type (all coated)	*Power (W)*	*Luminous Flux (lm)*	*Efficacy (LPW)*	*Color Temperature (K)*	*Color Rendering (CRI)*
HQL	125	6,300	50	4000	49
HQL DELUXE	125	6,500	52	3500	53
HQL SUPER DELUXE	125	5,700	46	3000	60

The life of high pressure mercury lamps is up to 24,000 hours The common wattages with luminous flux and efficacy are summarized in Table 8-2.

Self-Ballasted Lamps

Self-ballasted lamps are high pressure mercury lamps that incorporate inside of the outer bulb a current-limiting incandescent filament

Table 8-2. Data of important high pressure mercury lamps: luminous flux, luminous efficacy.

Power (W)	50	130	175	250	400	700	1000
Luminous Flux (klm)	1.8	4.5	8.9	13	22	40	58
Efficacy (LPW)	36	45	50	52	55	57	58

that is electrically connected in series with the mercury arc tube. This design has the advantage of eliminating the need for a current limiting ballast—i.e., the lamp can be connected directly to the supply voltage without a choke. The principle is shown in Figure 8-5.

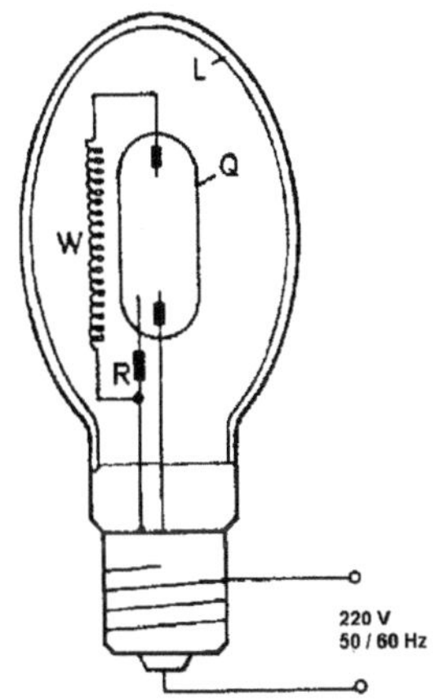

Figure 8-5. Schematic of self-ballasted high pressure mercury lamps: Q transparent quart glass discharge tube, W tungsten filament, R ignition resistor, L fluorescent coating.

Self-ballasted high pressure mercury lamps have, however, the disadvantage that the luminous efficacy is reduced by about 50% when compared with the standard high pressure mercury lamps. It is for this reason that self-ballasted lamps are only used as substitutions for incandescent lamp in fixtures which are difficult to access for lamp replacement. The life of this type of lamp, which is essentially controlled by the life of the incandescent filaments, is therefore rated for only 6000 hours.[13] The commercial assortment of such lamps and their luminous flux and efficacy data are listed in Table 8-3.

Table 8-3. Data of important self-ballasted lamps: luminous flux, luminous efficacy.

Power (W)	160	250	500	1000
Luminous Flux (klm)	3,1	5,6	14	32,5
Efficacy (LPW)	19	22	28	33

Metal Halide Lamps

Metal halide lamps are high pressure mercury lamps with metal halide additives. In general, most metal halides have higher vapor pressures than pure metals. The metal halides decompose in the core of the discharge and the metals can be excited and emit visible radiation. The intensity and spectral distribution depend on the kind of metal halides used, on their vapor pressure and on the core temperature in the arc. The emitted light fills the gaps in the mercury spectrum. Thus, the metal halides increase the luminous efficacy and improve the color rendering characteristics of the mercury discharge without the need to employ color-correcting phosphor coatings.

The first lamps were developed in the '60s. At that time, iodides of sodium (Na), thallium (Tl) and indium (In) were used.[11,12,13,14,15,16] Later, scandium (Sc)[17,18] iodides and halides of various rare earth metals were introduced such as dysprosium (Dy), holmium (Ho) and thulium (Tm).[19,20,21,22] and complex compounds with tin (Sn) and sodium (Na) halide.[23,24,25,26]

Investigations and discussions about additives have been published for almost all elements of the periodic table. However, there exists still an enormous potential for other compounds such as combinations of metals and the halides I, Br, Cl, F. This holds also true for the less investigated complex compounds of the metal halides.

Design

In the sixties, during the first years of the development, all metal halide lamps had elliptical, cylindrical or reflector-shaped outer bulbs and Edison screw bases like the mercury lamps on the market. Wattages were between 250W and 3500W (see Figure 8-6, a-d).[20,21] They were used principally in outdoor lighting, and in industrial, stage and sports stadium lighting. Later, smaller lamps, the double ended lamps in wattage of 400W, 250W, 150W and 70W were developed and also used for industrial interior lighting and display window lighting (see Figure 8-6, e-f).[21,27]

In the '80s, single-ended lamps in 150W, 70W and 35W with single pinched discharge tubes were added (Figure 8-6, g).

Arc Tubes

The double-ended MH arc tube is similar to that of the high pressure mercury lamps (see Figure 8-7).

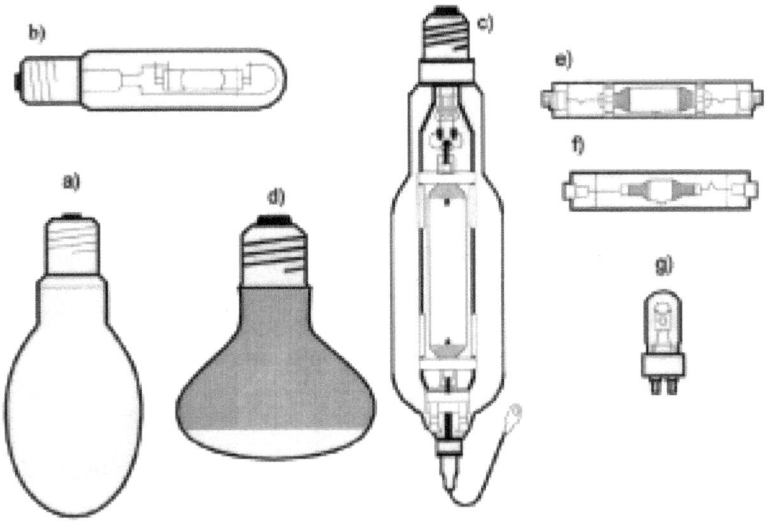

Figure 8-6. Metal halide lamps of various designs.

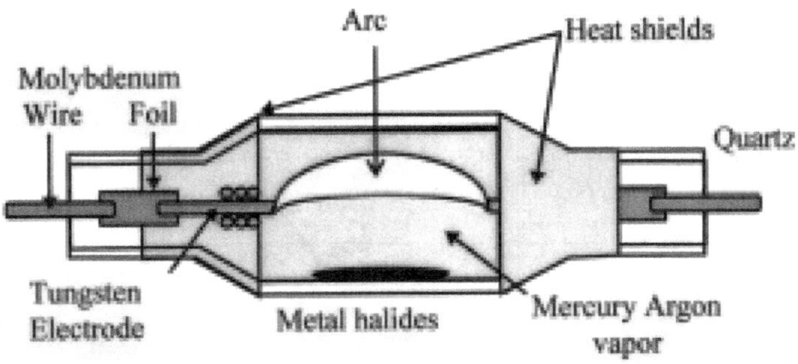

Figure 8-7. Arc tube of a metal halide lamp.

During operation, the metal halides are not completely evaporated. Therefore, a condensate is always located on the coldest spot of the discharge tube and it determines the halide vapor pressure and thereby the photometric and electrical data of the lamp. For this reason, MH arc tubes have to be formed with the objective of making the inner surface as isothermal as possible. By special forming of the discharge tubes, special design of the heat shields and defined electrode positioning the requirement of an isothermal inner surface is achieved (see Figure 8-8).

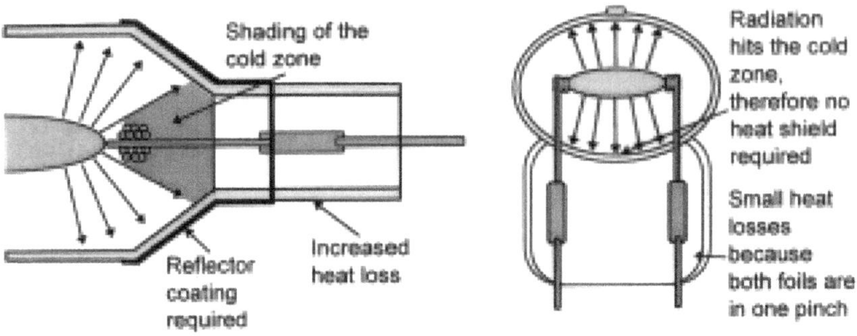

Figure 8-8. Comparison of double- and single-ended arc tubes.

Metal Halides

To obtain different color temperatures and meet color rendering requirements, different metal halide compositions are used. Representative data of a variety of commercially available lamps are listed in Table 8-4.

Table 8-4. Effects of various metals of halogen additives on MH lamp characteristics: W power, CCT color temperature, CRI color rendering, LPW luminous efficacy.

Halides:	*Na TI In*	*Na Sc Th*	*Dy Ho Tm Cs TI*	*Dy Ho Tm Na TI*	*Na Sn TI In Li*
Power (W)	250	175	250	150	150
CCT (K)	4800	3900	5200	4300	3000
CRI	60	70	93	85	76
Efficacy (LPW)	75	90	72	75	67

Better color rendering properties, but with lower lumen efficacies, are obtained by using the rare earth halides. To get cool white colors, sodium halides have to be added. A low color temperature of about 3000K can be obtained by using the sodium/tin halides.

Three typical spectral distributions of lamps with Daylight, Cool White and Warm White color are shown in Figure 8-9.

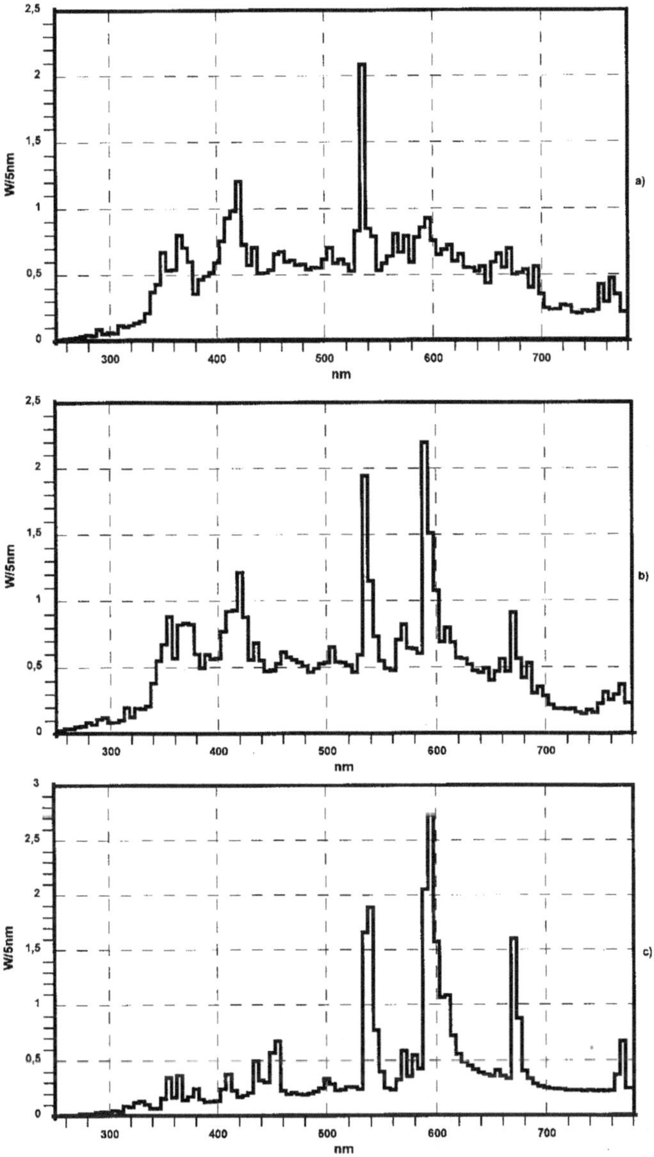

Figure 8-9. Spectral energy distribution of metal halide lamps with various colors: a) Daylight, CCT = 5600K, CRI = 90, additive elements Dy, Ho, Tm, Tl, Cs, Hg; b) Cool White, CCT = 4300K, CRI = 85, additive elements Dy, Ho, Tm, Tl, Na, Hg; c) Warm White, CCT = 3000K, CRI = 76, additive elements Na, Sn, Tl, In, Hg.

Complex Compounds

Some selected complex compounds evaporate undisassociated at the wall temperature of the arc tube. Therefore, the particle density of coveted elements—e.g., that of dysprosium can be increased in the arc because the complex compound $NaDyI_4$ has a higher vapor pressure than the DyI_3 itself. The dependencies of the partial pressures on the temperatures are reproduced in Figure 8-10.

The $NaDyI_4$ complex has a vapor pressure which is eight times higher at 1000 K compared to the pressure of the DyI_3. The investigation and application of complex compounds allows a relatively high particle density of the wanted elements in the arc at relatively small wall temperatures of the arc tube. High luminous efficacies and good color rendition properties are obtained.[27,28] The principle of this vapor pressure enhancement is used in European metal halide lamps having cool white color.

Ballasts and Ignitors

Conventional ballasts such as chokes and lead peak transformers were initially the ballasts of choice to limit the lamp current. Where necessary, ignitors have to be used in the circuits to start the lamps. Typical circuits are represented in Figure 8-11. These devices are relatively large and heavy and are a handicap to the introduction of small metal halide lamps. Therefore, small electronic ballasts with reduced weight have been developed operating the lamps at higher frequencies than 50 or 60 Hz.

In comparison to the conventional ballasts, electronic devices have the following advantages:[29,30,31]

- Small weight and volume;
- Higher system efficacy;
- Integrated striking equipment;
- Constant power control regardless of the supply voltage and of the lamp voltage, resulting in less color variation;
- Less flickering effects; and
- Smaller arc bending.

The difficulty with the development of such devices is the choice of a suitable frequency and the curve shape of the supply voltage. With

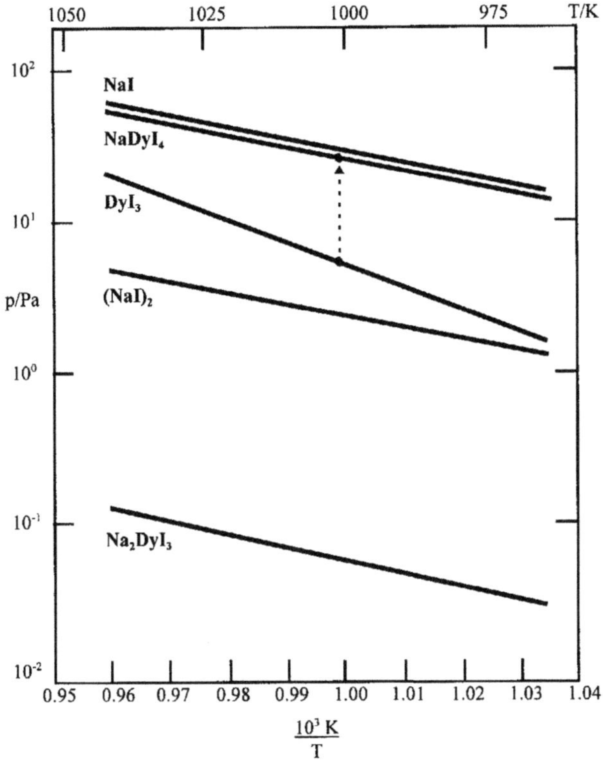

Figure 8-10. Partial pressures of molecules NaI, $NaDyI_4$, DyI_3, $(NaI)_2$ and Na_2DyI_3 of a mixture of 50 mol% NaI and DyI_3.

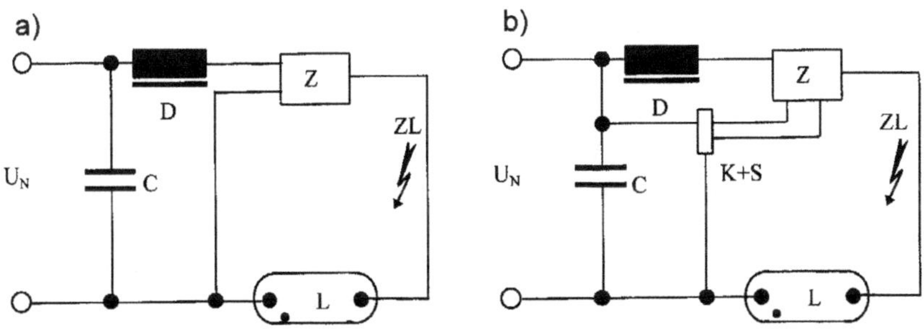

Figure 8-11. Circuits for MH lamps with conventional ballasts: a) for normal ignition, b) for hot restarting; D choke, C capacitor for power factor correction, K + S short term switch and contact, L lamp, UN supply voltage, Z ignitor, ZL high frequency ignition line.

many frequencies, high pressure discharge lamps operate in an unstable mode or the arc extinguishes due to acoustic resonances which are produced in the plasma. The resonance frequency is strongly dependent on the arc tube size, arc tube shape and the filling materials.[7,29] Power levels and the most important data of several MH lamps are shown in Table 8-5.

Table 8-5. Summary of important metal halide lamp data.

Type Power (W)	35	70	150	250	400	1000	2000	3500
Warm White								
single-pinched arc tube								
lumin. flux (klm)	2.4	5.2	12.0					
efficacy (LPW)	62	75	80					
Cool White								
lumin. flux (klm)		5,5	11,3	20				
efficacy (LPW)		73	75	80				
Daylight								
lumin. flux (klm)				19	25	90	170	300
efficacy (LPW)				76	69	90	85	86

Lamps for Special Applications

Medium Short-Arc Metal Halide Lamps

MH lamps of even higher luminous efficacies and improved color properties are obtained by increasing the wall loading, but at the expense of a reduction in lamp life. At higher operating temperatures, the filling substances react strongly with the materials of the arc tube—specifically, the quartz glass and the tungsten electrodes.

Life of 100-1000 hours is sufficient for many lighting applications—e.g., for film and television studios as well as for film projection and endoscopy. Therefore, medium short-arc metal halide lamps of the type HMI have been developed, using higher wall loadings, shorter arc lengths, and with higher brightness than metal halide lamps that are used for general lighting.

Because of the high wall loading, no outer bulb is required. In order to be able to use the lamps for film and TV recording, a good color rendition and a color temperature of 5600K are obligatory. Those characteristics are realized with the halogens of the rare earth metals dysprosium (Dy), holmium (Ho), thulium (Tm) and gadolinium (Gd).[32] The design of the lamps consists basically of a quartz glass arc tube as represented in Figure 8-12. A typical spectral distribution is shown in Figure 8-13.

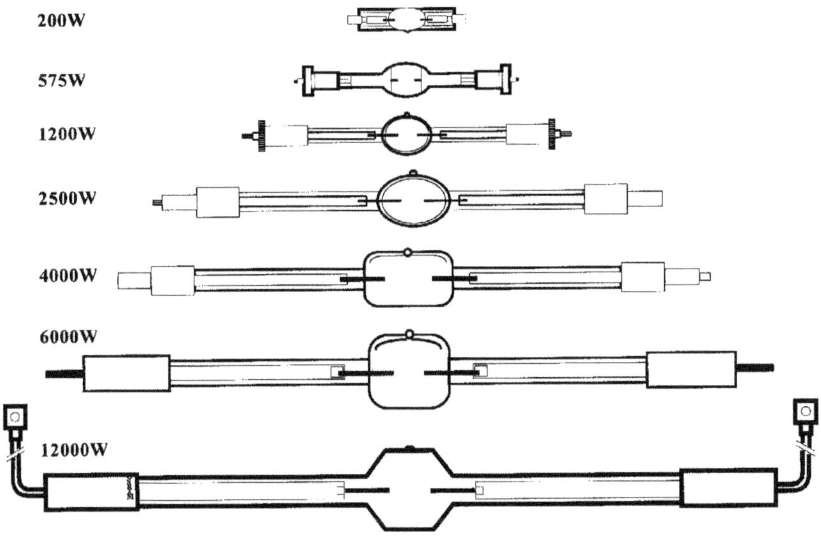

Figure 8-12. Design configurations of medium short-arc metal halide lamps: 200W to 12000W.

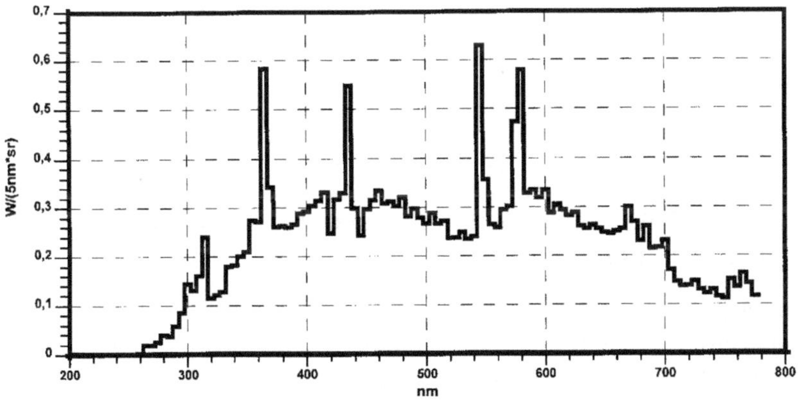

Figure 8-13. Spectral energy distribution of a 1200W medium short-arc metal halide lamp.

The lamps are operated in projectors with chokes and special ignitors that can reignite even hot lamps. Often electronic ballasts are used. The lamps are available in wattages from 200W to 12kW.[33,34,35] Important data are shown in Table 8-6.

For use in small projectors, often as replacement for similarly shaped incandescent lamps, single-ended HM lamps with outer bulb were developed (see Figure 8-14 and Table 8-7).

Table 8-6. Data of HMI metal halide lamps, Daylight 5600K, color rendering CRI > 90.

Power (W)	200	250	575	1200	2500	4000	6000	12000
Lumin. Flux (klm)	16	16.2	49	100	240	413	570	1100
Efficacy (LPW)	80	65	85	88	96	102	95	92
Brightness (kcd/cm^2)	40	10	10	12	12	11	17	23

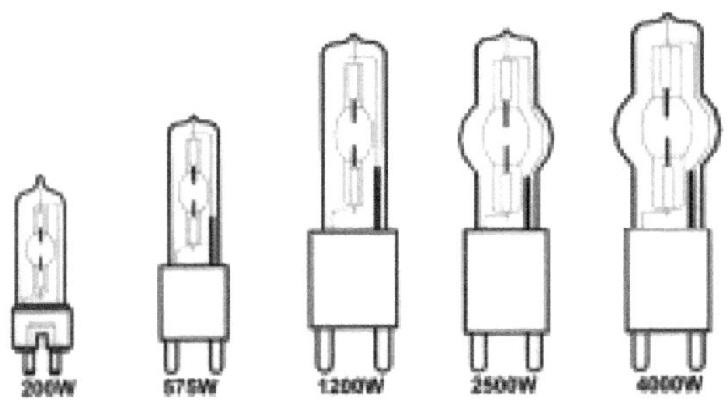

Figure 8-14. Design configurations of single-ended medium short-arc metal halide lamps with outer bulb.

Table 8-7. Data of single-ended medium short-arc metal halide lamps with outer bulbs.

Power (W)	200	575	1200	2500	4000
Luminous Flux (klm)	16	49	110	240	380
Efficacy (LPW)	80	82.5	91.6	96	95.2
Color Temperature (K)	6000	6000	6000	6000	6000
Base	GzY 9.5	G22	G38	G38	G38

Short-arc Metal Halide Lamps

Metal halide lamps of the type HTI with even shorter electrode spacings are used for optical purposes. Two types have been developed with an integrated dichroic reflector for use in medical and technological applications (endoscopy).

The HTI short-arc lamps are characterized through high luminous efficacy and high brightness and are used in projection light sources and for special light effects. The filling materials are similar to those of the HMI lamps. Their design is shown in Figure 8-15.

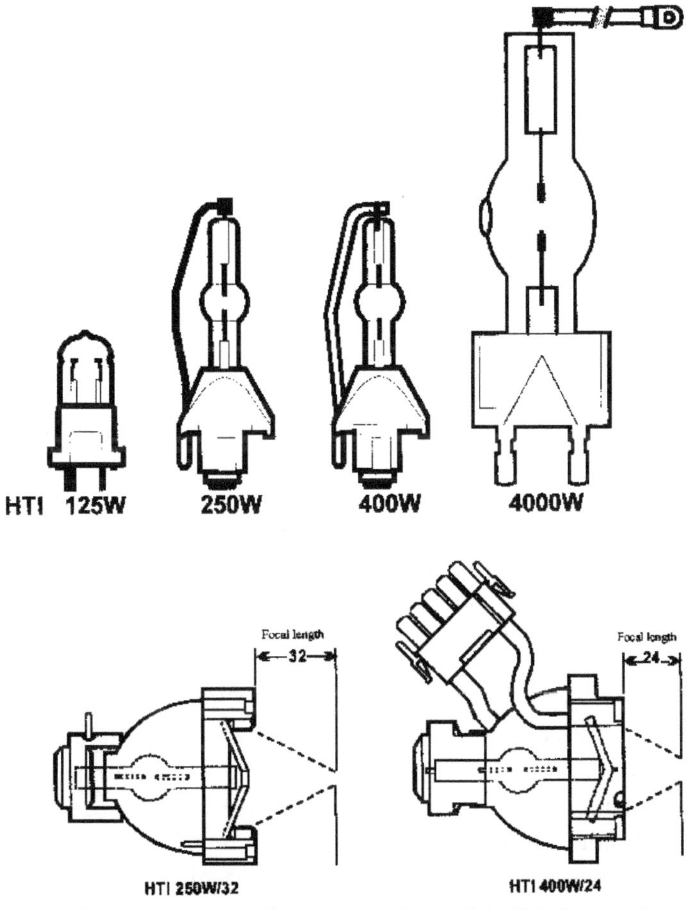

Figure 8-15. Construction and wattage of metal halide lamps for optical applications.

The dichroic reflector is transparent for heat radiation and reflects the visible light only in the forward direction. Therefore, the thermal stress of illuminated objects is strongly reduced. Figure 8-16 shows the spectral energy distribution of the arc tube: a) the reflected, b) the transmitted and c) the backward-radiated. The most important data are presented[36,37] in Table 8-8.

Table 8-8. Important data of single-ended short-arc metal halide lamps for optical applications.

Power (W)	150	250	400	1200	2500
Luminous Flux (lm)	10000	20000	30000	110000	220000
Brightness (kcd/cm^2)	5	40	30	26	30
Color Temperature (K)	6500	4600	4600	6000	6000

UV Sources

The high pressure mercury discharge (HPM) typically radiates about 15% of the power consumption as ultraviolet (UV) radiation. Of this power, 6% is radiated in the UV-A (400 to 315 nm), 5% in the UV-B (315 to 280 nm) and 4% in the UV-C (280 to 200 nm) region of the spectrum.

UV radiation is needed for many technical, medical and cosmetic applications. For those purposes, the UV radiation can be increased by using special metal halide additives.[38,39] Halides of the ferromagnetic metals Fe, Co and Ni increase the radiant efficiency by approximately 20% of the power consumption, mostly in the UVA region.

The UV radiators are similar in design to standard metal halide lamps, except that their arc tubes must operate with higher wall loadings. Therefore, their arc tubes are smaller and the lamps are mostly designed without an outer bulb (see Figure 8-17). The spectral energy distribution for a 400W lamp of this type is reproduced in Figure 8-18. Lamps of this type can be used for the treatment of psoriasis or for UV curing of paints and resins.

Important data of UV sources are summarized in Table 8-9.

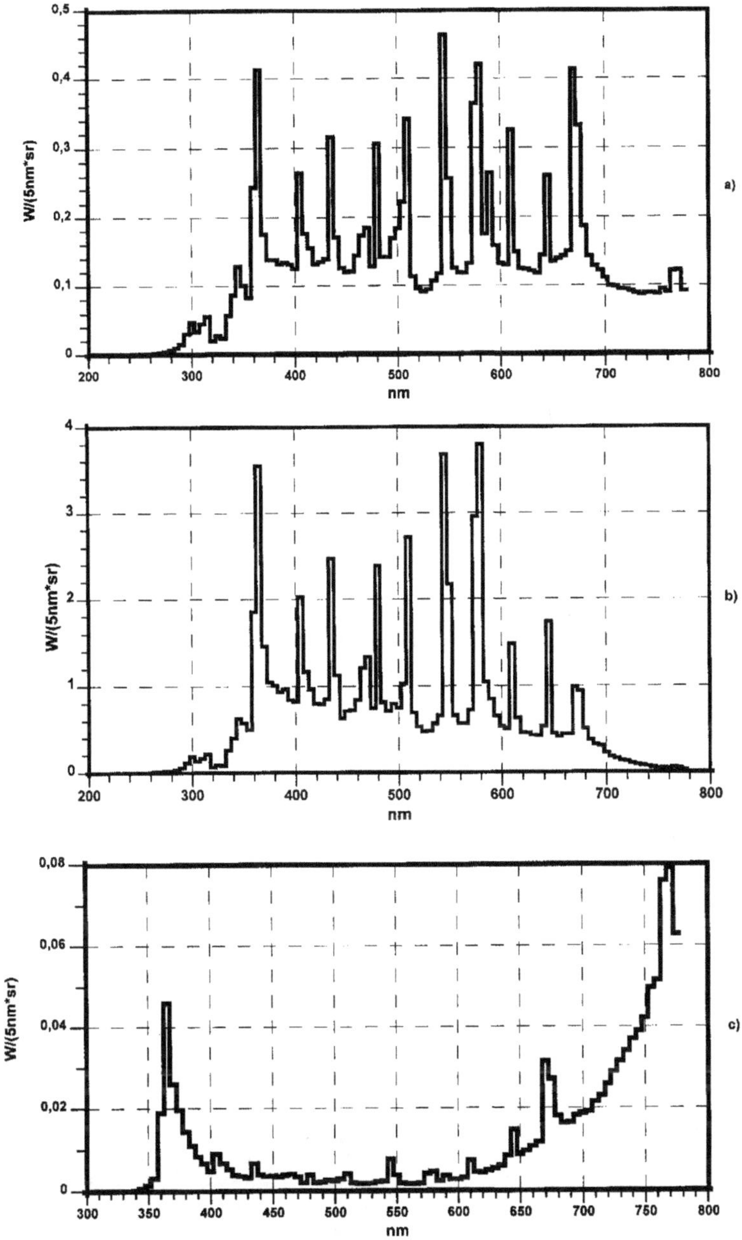

Figure 8-16. Spectral energy distribution of a 400W metal halide lamp with dichroic reflector: a) radiation of the arc tube, b) forward reflected radiation and c) backward-transmitted radiation.

Table 8-9. Data for UV radiators.

Power (W)	400	1000	2000	3000	4000
UV Radiant Power (W)	88	260	600	900	1200
UV-A (315-400 nm) (W)	75	200	400	600	800

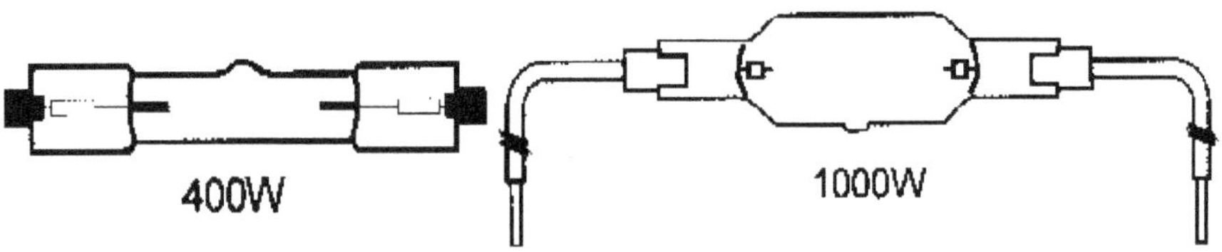

Figure 8-17. UV radiators are tubes similar to high pressure mercury arc tubes.

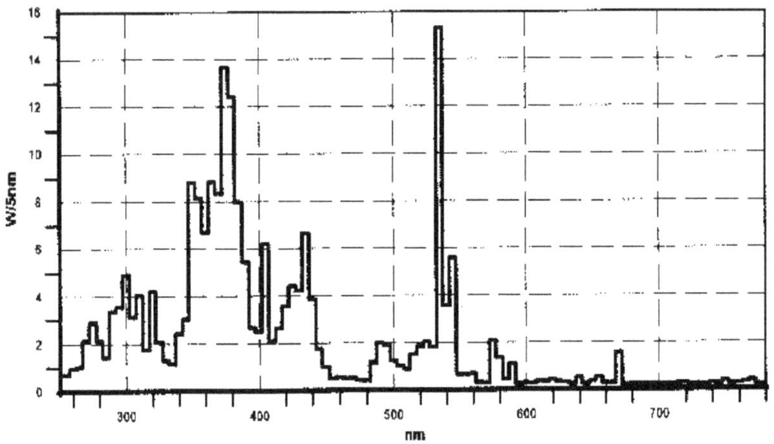

Figure 8-18. Spectral energy distribution of a 400W UV enhanced radiator.

More Recent Metal Halide Lamp Developments

In the past 13 years, the development of a variety of new lamp designs for special applications has been possible due to the progress in production technologies and in the purity of the arc tube, the metals and the dosing materials. Floodlighting was one of the first areas where the efficacy was in the range of 130 LPW without significant reduction in lamp life.

Another advance was made by miniaturizing the arc tubes and solving the problem of warm-up time, which made it possible to use MH discharge lamps for automotive head lamps.

One of the most challenging areas was the ceramic technology where success was achieved in solving the sealing problems encountered with ceramic MH arc tubes.

Add to this electrodeless discharge lamps which allow the use of chemically aggressive halide chemicals and offer improved luminous efficacy, better color rendering and longer life.

Floodlighting

One of the most critical temperature affected areas for HID lamps is the base where the molybdenum foil comes into contact with the air. Temperatures above 400°C shorten lamp life by oxidizing the molybdenum foils. By further improving the pinch seal technology, it was possible to lower the temperatures at the critical pinch seal area to a level which allows the lamp to be operated without the protective outer bulb reaching the specified life. This permits the desired compact structural design for projectors and, moreover, eliminates the absorption losses caused by the outer bulb design.

Both lamps use elliptical discharge tubes with identical pinch seal and basing. Ceramic bushings are used as lamp holders and nickel stranded wires with cable shoes provide safe contacts. By the two-sided lead-in design, an instantaneous hot restart is possible when the ignitor provides the required peak voltage of 36kV.

The new design is shown in Figure 8-19.

In Table 8-10, the older version of a HQI-TS 2000W/D lamp is compared with the new designed HQI-TS 1000W and 2000W/D/S lamps without outer bulb.

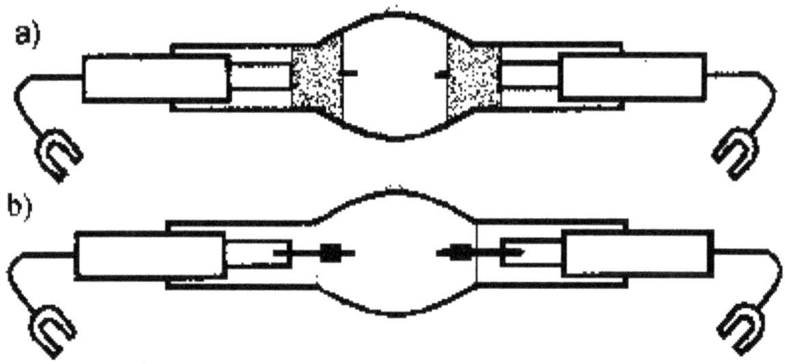

Figure 8-19. New lamps for floodlighting without outer bulb: a) 1000W and b) 2000W.

With the new lamp design, smaller reflectors can be used, thereby reducing the wind load on the light pole.

Through these changes, a new generation of ideal light sources has been developed for economical sports stadium lighting with about a 15% higher light system efficiency.

Table 8-10. Technical data for high wattage HID lamps for stadium lighting.

Type	 HQI-TS 2000W/D	 HQI-TS 2000W/D/S	 HQI-TS 1000W/D/S	
Supply Voltage	380	380	220	Volt
Lamp Voltage	215	215	115	Volt
Lamp Current	10.3	10.3	9.5	Ampere
Luminous Flux	170000	200000	95000	lumens
Electrode Gap	105	31	31	mm
Brightness	1300	6600	3300	cd/cm^2
Efficacy	85	100	95	LPW
System Efficacy	82	97	92	LPW
Useful Lamp Life	750	2500	6000	hours

Automotive Head Lamp Systems

There are two trends in the development of modern automobile headlights:

1. The demand for optimum vehicle aerodynamics, more compact headlights, and the desire for an integrated design is leaving headlight designers space for only small headlight systems.

2. The call for brighter and better distributed light for the best possible view of the road ahead and the need for more safety in nighttime driving are placing new and greater demands on the headlight system.

Halogen lamp technology, which was revolutionary more than two decades ago, has reached its limitations and cannot meet the objective to generate "more light from a smaller headlight."

Gas discharge technology, on the other hand, offers a number of advantages, such as higher luminous efficacy, reduced heat radiation and longer lamp life. These benefits have long been the practice in general lighting and technical applications. A lamp based on these principles was developed by Osram as part of a European project. The light distribution of this lamp is compared with that of a TH lamp in Figure 8-20; its construction is shown in Figure 8-21. The lamp draws 35 watts and has a discharge tube of only 0.03 cm^3. The color temperature is 4000K compared to 3200K for the TH lamps of the types H1 and H2, a little whiter and closer to natural daylight.

The advantages of the gas discharge headlamps over the TH lamps are shown in Figure 8-22.[40] The lamp must be operated with a special electronic ballast that generates high run-up currents and shortens the warm-up period (see Figure 8-23). The TH headlamp reaches its maximum light output within 0.06 seconds. During this time, the gas discharge lamp emits light produced by the xenon fill gas. Upon warm-up of the metal halide additives, this is followed by a rapid rise to full light output after three seconds.

In normal use, the gas discharge headlamps have a life of 2,000 hours, whereas the conventional TH headlamps generally need to be replaced more than once. The technical data of the two headlamp systems are compared in Table 8-11.

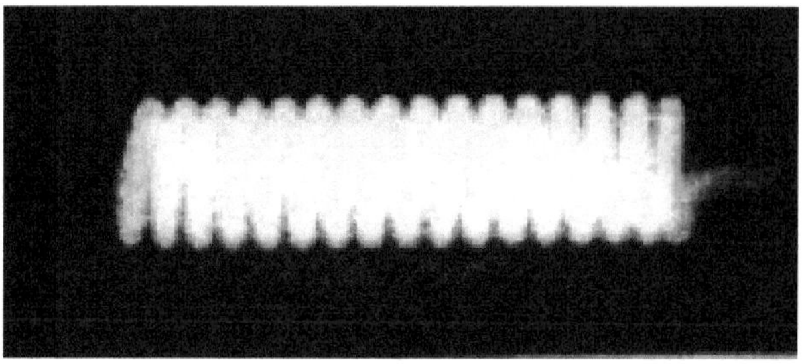

Conventional halogen lamps in car headlights generate light by passing an electrical current through a tungsten filament to make it glow white.

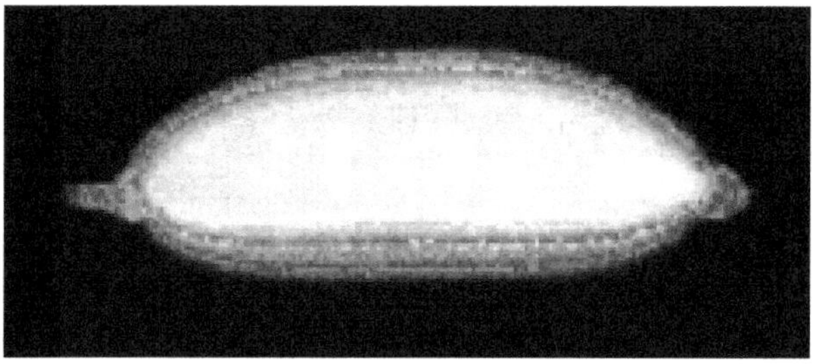

The gas discharge lamp does not have a filament. Instead, an arc is created between two tungsten electrodes in an extremely small quartz bulb. The bulb is filled with xenon gas and metal halides.

Figure 8-20. Shape of the light emitted from a filament vs. shape of the light emitted from an arc lamp.

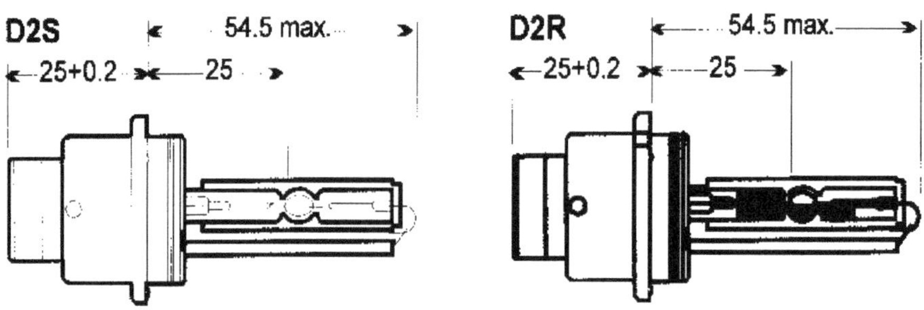

Figure 8-21. Two 35W automotive headlight assemblies using gas discharge lamps. Courtesy of OSRAM.

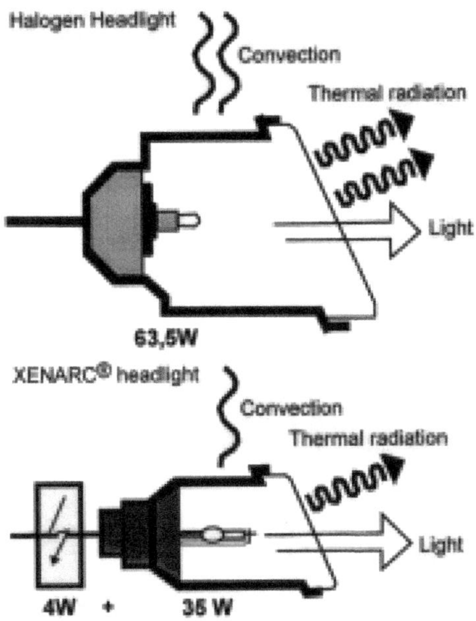

Figure 8-22. Comparison between the tungsten halogen and gas discharge headlights.

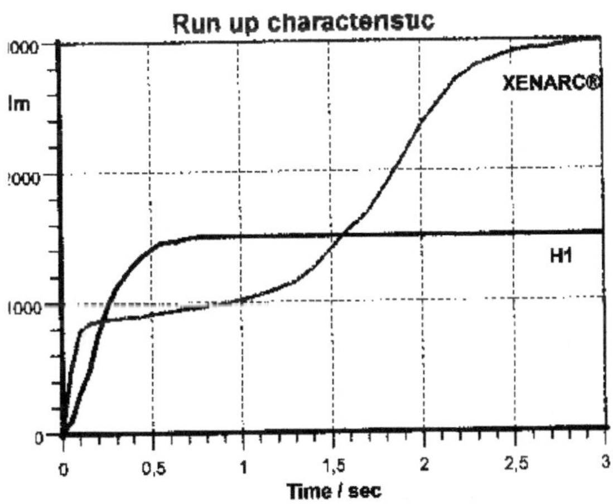

Figure 8-23. Luminous flux of incandescent H1 and gas discharge D2S lamps during run-up.

Table 8-11. Technical data for automotive headlights: comparison of TH (H1) versus xenon gas discharge lamps (Osram models D2S, D2R).

	H1	D2S	D2R
Lamp Wattage (W)	63.5	35±3	35±3
Lamp Voltage (V)	12±20%	85±20%	85±20%
Luminous Flux (lm)			
After 1 second	1200	800	700
After 60 seconds	1500±15%	3200±15%	2800±15%
Color Temperature (K)	3100	4250	4100
Average Lamp Life (hrs)	500	2000*	2000*
Base		P32d-2	P32d-3

*For a 120-minute switching cycle.

Ceramic Arc Tubes for Metal Halide Lamps

Limits of Quartz Technology

The types of halides that offer good photometric properties cannot be used at the preferred high temperature in discharge tubes made of quartz, due to various undesirable chemical reactions of the metal halide additives with the quartz glass arc tube. These mechanisms lead to considerable color shift over the life of the lamp. Moreover, depletion of sodium also causes considerable lamp voltage rise, unstable operation and shortened life.

These reactions of halides with quartz do not occur with ceramic arc tubes of alumina. Analogous to quartz, rare earth elements do react with alumina to form aluminates and aluminum halide. However, in contrast to silicon, aluminum hardly dissolves in tungsten, and so the reaction is self-limiting. Therefore, there is minimal depletion of the metallic parts of the halide additives (see Figure 8-24). In addition, transport reactions on the electrodes are minimized and therefore, a stable operation over time is assured.

Permeation of sodium ions through the alumina tube wall is negligible, in comparison to that through the quartz arc tubes. Elimination of both depletion mechanisms guarantees a stable halide composition, and thus a small initial color spread, little variation from lamp to lamp and an excellent color stability over life.

This chemical stability of alumina allows the discharge tubes to be operated at a higher temperature than quartz. It is difficult to provide a meaningful comparison of the surface temperature profiles of quartz and alumina discharge tubes, but a comparison of the cold-spot temperatures does provide some insight. Typical cold-spot temperatures of alumina discharge-tubes are 950-1050°C, while those of quartz are lower at 800-900°C.

Alumina, as a material for the discharge-tubes of metal halide lamps, does not show a huge difference in its chemical properties but more in its thermal properties (see Table 8-12).

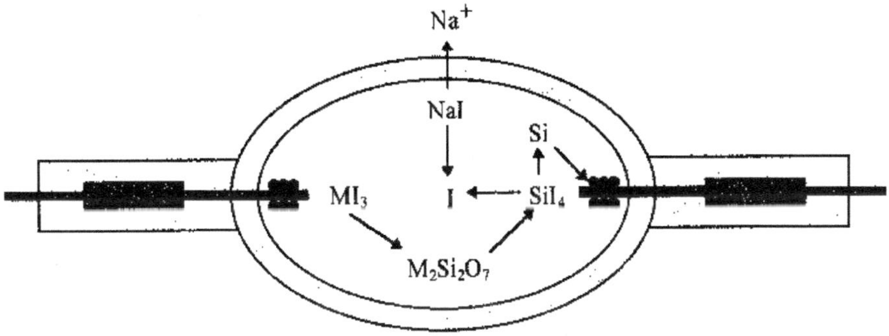

Figure 8-24. Reactions in MH-quartz lamps.

Table 8-12. Thermal properties of quartz and alumina at 1200K.[41,42,43]

	Quartz	*Alumina*
Thermal Conductivity (W/mK)	4	12
Thermal Emission Coefficient	0.35-0.55	0.15-0.35
Thermal Expansion K^{-1}	$4*10^{-7}$	$8*10^{-8}$

These differences have two consequences:

First, the thermal distributions in the arc-tube walls are very different, and so the arc tubes must have different configurations, in order to realize the required cold-spot temperature. Second, realizing reliable electric feed-throughs requires different materials and constructions, because the halide-resistant lead-through material molybdenum, that is used in quartz glass arc-tubes, has a coefficient of thermal expansion smaller than that of alumina. This can cause the formation of cracks in the alumina.

To take full advantage of the potential that alumina offers, the arc tube needs a complete redesign as it is shown in Figure 8-25.[44]

With the extended plugs, the temperature of the critical zones, where the sealing takes place, can be lowered below the critical values. The frit material serves also as a protection for the niobium wires against the chemical attack from the metal halides.

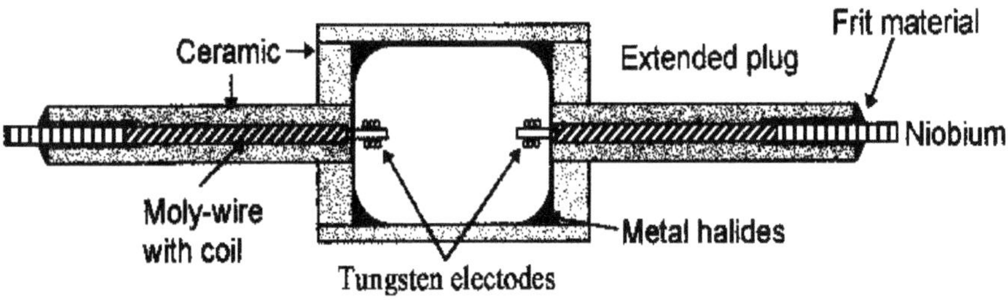

Figure 8-25. Ceramic arc tube design.

Properties of Metal Halide Lamps with Ceramic Arc Tubes

Metal halide lamps with ceramic arc tubes offer interesting photometric properties. These properties provide lamps with improved color rendering, color control and luminous efficacy, both initially and throughout lifetime. All improvements can best be illustrated by a comparison of three types of metal halide lamps: quartz arc tubes with sodium scandium fill, quartz arc tubes with sodium/rare-earth fill, and alumina arc tubes with sodium/rare-earth fill, shown in Table 8-13.

Color Rendering

The differences in color rendering among the three types of metal halide lamps is reflected by their spectra. The lamps with alumina arc tubes show a higher continuum radiation throughout the spectrum, and especially in the red part of the spectrum.

Table 8-13. Typical properties of 70W and 100W metal halide lamps. Data of color temperature, CRI, and luminous efficacy indicate catalog values of lamps of major lamp manufacturers. Data of color spread indicate 3-sigma values of measurements of at least 10 lamps.[45]

Arc Tube Material	*Quartz*	*Quartz*	*Alumina*
Fill-System	*Sodium/ Scandium*	*Sodium/ Rare Earth*	*Sodium/ Rare Earth*
Color Temperature (K)	3000	3000	3000
Color Spread (lamp to lamp) (K)	360	250	150
Color Rendering (CRI)	65-75	78-82	82-85
Efficacy (LPW)	75-80	75-80	90-92

Color Control

The improved color control of lamps with ceramic envelopes is manifested as reduced color spread and improved color stability.

Luminous Efficacy and Lumen Maintenance

The luminous efficacy of lamps with ceramic envelopes is significantly higher than that of lamps with a quartz envelope, due to their higher operating temperature.

Electrodeless Microwave HID Lamps

The technology behind electrodeless discharge lamps (or more simply, electrodeless lamps) is over 130 years old: Nikola Tesla demonstrated the first electrodeless low pressure lamp in 1891. Because of advances in the electronics industry and changes in electromagnetic interference (EMI) standards over the last 30 years, electrodeless lamps are now starting to appear on the general lighting market.

All electrodeless lamps use an EM field, instead of an electric current passing through electrodes, to excite the gas fill in a bulb. The magnetron generates a microwave field which travels through a wave guide into a special cavity which serves as fixture.

Inside the resonance cavity, a quartz capsule is situated which contains a starting gas, mercury and the metal halides as the light emitting materials. If the load is very high, the bulb has to be rotated to get uniform light and to prevent the bulb from being overheated.

Since mid-1994, the sulfur lamp has appeared as a prototype. At 3130W of applied microwave power, 440,000 lumens were obtained with this prototype.[46]

A major drawback of that design was an additional forced cooling with jets of air, which produced acoustic noise. To get rid of this problem, the power was reduced to 1000W and the discharge tube was enlarged. Thus a system efficacy of about 99 LPW was obtained.

The sulfur lamp produces a large amount of light, making it convenient for remote source lighting applications. At the National Air and Space Museum (Smithsonian in Washington, DC, and also at the Forrestal Building in Washington, DC), three light pipes, each lighted at one end by a single sulfur lamp, replaced 94 HID lamps.

Long life is the most touted feature of electrodeless lamps. Because they have no filament or electrode, they do not fail from sputtering, breakage by shock or other electrode-related phenomena. The sulfur lamp's bulb has practically an infinite life, because the sulfur and argon in its fill do not react with each other or with the glass. Therefore, the life of the microwave generator limits the lamp system's life. Replacing the magnetron with solid-state technology can extend system life.

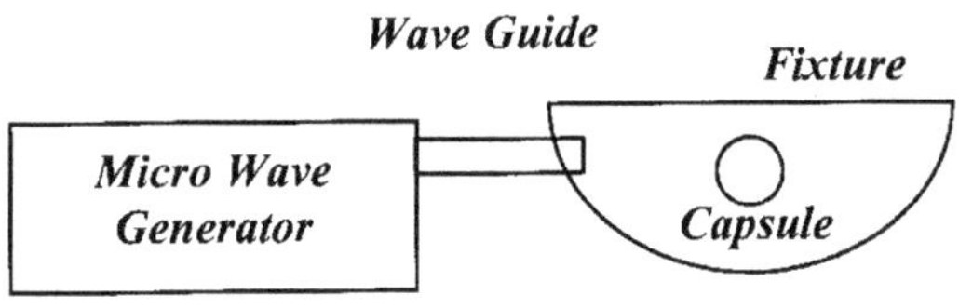

Figure 8-26. Principle of operation of the sulfur lamp system.

Ultra High Performance Projection Lamps[47]

In the last years TV projection systems using liquid crystal panels to generate the TV picture have become of increasing importance. The rapidly growing market depends strongly on the availability or suitable projection lamps with a color temperature of about 8000 K and a high luminance.

The first such high pressure short arc lamps where operated with metal halide filling. These lamps emit atomic line radiation from various metals and molecular bands from the corresponding halides. Due to the low ionization energy of the additives the plasma temperature in these lamps is between 5000 K and 6000 K.

Rising the power density by shortening the electrode gap shifts the radiation intensity closer to the black body limit, but this method is limited by the stability of the electrodes. Though metal halide lamps represent a major improvement compared with incandescent lamps, they are still far away from the ideal point source.

To create projection lamps with extremely high luminance the path directed by Elenbaas has been followed as a alternative way. Normally mercury vapor is regarded as an atomic gas.

The existence of diatomic molecules is well known, but the molecular ground state is only loosely bound, so that the concentration of Hg_2 is low at normal pressure. The spectrum emitted by mercury plasmas is dominated by atomic lines molecular band radiation is very weak. But as the molecule concentration rises with the square of pressure. The contribution of molecular band radiation becomes important, when the operating pressure is increased. The spectral distribution of such a ultra high pressure mercury lamp is shown in Figure 8-27.

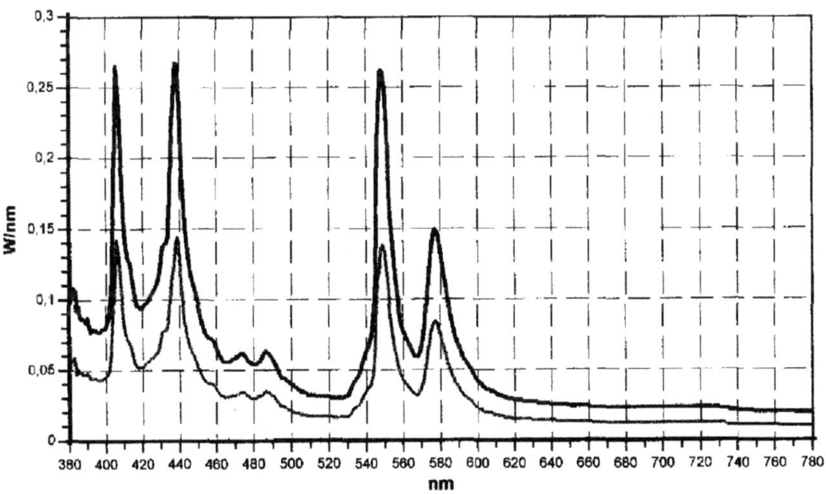

Figure 8-27. Spectral distribution of a UHP lamp operated at 120 W with different apertures—black curve 6mm—grey curve 3mm.

The lamp has an arc length of 1.3mm and operates at a pressure of about 200 bar. The arc bulb has a spherical shape, to avoid optical distortion of the emitted light bundle. Figure 8-28 shows the burner geometry. The cylindrical parts extending from the spherical bulb have been chosen very long to avoid thermal corrosion of the molybdenum foils. This is necessary, because the lamp is designed for operation on air without an outer bulb, to keep as much as possible of the high luminosity in the projected light beam. Any additional glass surface would lower the brightness due to reflection or scattering. The nominal operating voltage of the lamp is 85 V. A critical point of lamps with such small arc gaps is, that a small deviation from the nominal value leads to a large change of the arc voltage. A deviation by only 0.1 mm leads to a voltage change of about 6 V. When operating the lamp with a conventional inductive ballast, this would lead to a change of the consumed power by 15 %. This again would disturb the thermal balance or the lamp tube and thus lead to reduced lamp life or light quality in respect of luminance, luminous flux and color. Therefore a special electronic ballast with power stabilization has to be used, to keep the lamp in optimal operating conditions.

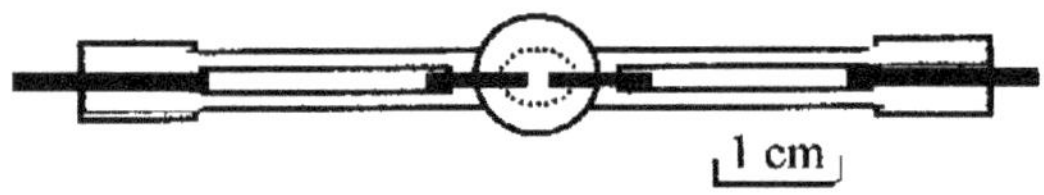

Figure 8-28: UHP 120W-Arc Tube

Due to the low arc current, the thermal load of the electrodes is low and consequently also the evaporation of the tungsten electrodes. But due to the small inner surface even this rate would lead to severe blackening of the tube wall. The absorption of radiation would strongly influence the thermal balance of the wall. By adding an appropriate bromine content to the fill, a circle process can be established which prevents the deposition of tungsten at the wall.

As there is no wall blackening and no chemical reactions, which could change the plasma composition, the color rendering properties are nearly constant during life.

Future of Metal Halide Lamps

The first step into the future will be development of lamps with smaller wattages and designs. Such small discharge lamps can be used in all the other fields of general lighting. Other possibilities of improving the lamps in the areas of luminous efficacy and color rendering have been shown in a 150W lamp with Dy, Ho, Tm, Na, Tl halides. These lamps were measured in wattages between 95W and 350W, giving higher luminous efficacies and better color properties with higher wall loadings (see Figure 8-29). However, chemical reactions of the filling material with the quartz glass shorten the life of such lamps, or exceed the softening point of the quartz glass.

With Na, Sc halides, the 100 LPW limit can be easily exceeded with increased wall loading, however the color rendering is better with rare earth halides.

These examples show that the MH lamp development has to be continued with the objective of finding materials that tolerate higher loadings than quartz glass. The Al_20_3 ceramics already used for some types, as described before, will also be used in the future for higher wattages.

Further improvement of MH lamps becomes evident when one looks at the energy distribution of the total radiation of the MH lamps in Figure 8-30. The visible portion of the entire radiation is only 20-40%, with other major portions emitted in the UV and the infrared. By use of suitable coatings or doping of the arc tube and outer bulb with IR and UV reflecting or absorbing materials, the radiation losses can be reduced

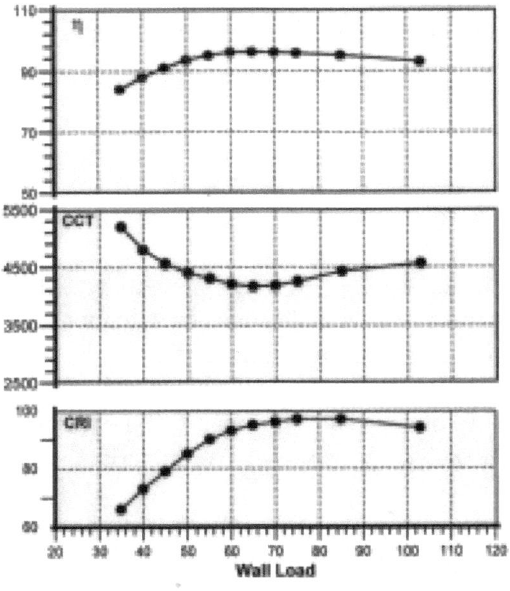

Figure 8-29. Luminous efficacy LPW, color temperature CCT and color rendering CRI of overloaded metal halide lamps with single-ended tubes, filling materials: Dy, Ho, Tm, Tl - halides + Hg.

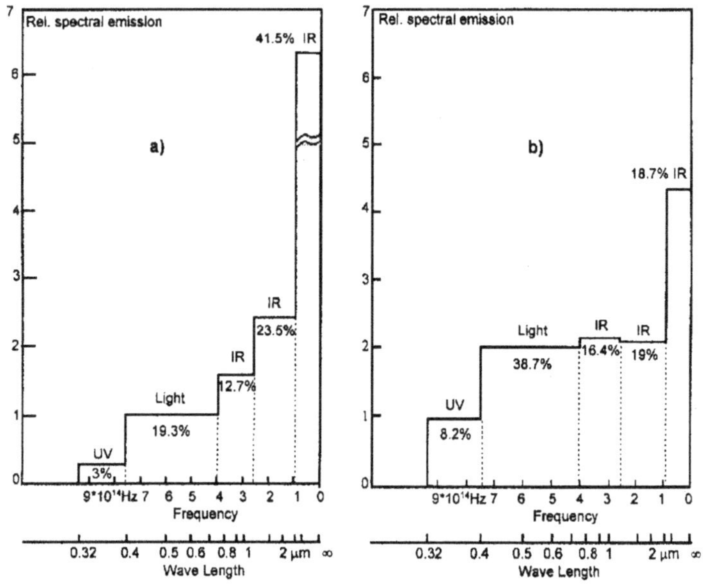

Figure 8-30. Total radiant power of metal halide lamps: a) HQI-TS 70 W/WDL with Na, Sn, Tl, In, halides + Hg, b) HQI-TS 250 W/NDL with Dy, Ho, Tm, Na, Tl, halides + Hg.

and the luminous efficacy increased.

Conclusion: The most important development steps for the future of the metal halide lamps will be:

- Low wattages (20W to 150W);
- Different color temperatures (3000K to 5600K);
- Improved color rendition;
- Higher efficacy and improvements in life by using higher vapor pressures and better complex-forming compounds;
- Reduction of IR losses by coatings and/or dopants of the lamp vessels with UV-and/or IR-reflecting or -absorbing materials;
- Use of electronic ballasts with power control and instant ignition electrodeless MH-discharges.[46,48,49,50]

References

1 A. Dobrusskin, Handbuch der Vakuum Elektronik, Eichmeyer/ Heynisch(Hrg.) Oldenbourg Verlag Munchen Wien 1989 S. 538-569
2 W. Elenbaas, The high pressure mercury vapor discharge, North-Holland Publ. Comp. 1951
3 H.P. Stormber, R. Schafer, J. Appl. Phys. 54, 4338, 1983
4 R.J. Zollweg, J. Appl. Phys. 49, 1377, 1978
5 E. Fischer, J. Appl. Phys. 47, 2954, 1976
6 C.H. Sturm, Brown, Boverie & Cie AG, Mannheim Vlg. W. Girardet, Essen, 1974
7 E. Statnic, Techn. Wissensch. Abhandlungen der OSRAM-Gesellschaft Vol. 12, p. 394, 1986
8 H.J. Fähnrich, E. Rasch, IES-Anual Conference 1987
9 E. Rasch, Lighting Design + Application Vol. 17, No. 9, 1987
13 Dobrusskin, B. Kühl, Lichttechnik 15, H. 11, 1963
11 B. Kühl, H. Krense, DDP 1184008, 12.8.1960
12 D.A. Larson, H.D. Fraser et al. *J. Ilum. Eng. Soc.* 58, 1963, No. 6, pp. 434-439
13 E.C. Martt, L.J. Smialek et al. *J. Ilum. Eng. Soc.* 59, 1964, No. 1, pp. 34-42
14 G.H. Reiling, J. Opt. Soc. of America 54, 1964, No. 4, pp. 532-540
15 B. Kühl, Lichttechnik 16, 1966, No. 4, pp. 68-71
16 L.B. Beijer, C.A.J. Jacobs, T. Tol. Philips Techn. Rdsch. 29, 1968, No. 11/12, pp. 345-354
17 A. Bauer, Lichttechnik 16(1964), No. 3, pp. 118-120
18 Koury and Waymouth, U.S. Patent No. 3407327, 1968
19 F. Rokosz, W.J. Decker, H.D. Fraser, National Technical Conference of the Illuminating Engineering Society, Montreal, Canada, Sept. 1967
20 A. Dobrusskin, Lighting Research and Technology, Vol. 3, 1971, No. 2, pp. 125 - 132
21 A. Dobrusskin, H. Leyendecker, TechnischWissenschaftliche Abhandlungen der OSRAM-Gesellschaft, Vol. 11, 1972, pp. 163-188
22 A.G. Chalmers, D.O. Wharmby, F.L. Whittaker, Lighting Research and Technology Vol. 7, 1975, No. 1, pp. 11-18
23 R. Lorentz, International Symposium on Incoherent Light Sources. Loughborou April 1977
24 A. Dobrusskin, Lichttechnische Gemeinschaftstagung Mannheim, 1984
25 D.G. Fromm, J. Seehawer, I.W. Wagner, B. Winzerling, German Patent AppJn. 1 2655167.7, filed Dec. 6, 1976
26 D.G. Fromm, J. Seehawer, I.W. Wagner, Lighting Research and

Technology, Vol. 1, 1979.
27 A. Dobrusskin, D.G. Fromm, J. Heider, TechnischWissenschaftliche Abhandlungen der OSRAM-Gesellschaft. Vol. 12, 1986, pp. 11-30
28 K. Hilpert, J. Seehawer, TechnischWissenschaftliche Abhandlungen der OSRAM-Gesellschaft, Vol. 12, 1986
29 J.M. Davenport, R.J. Petty, *J. Illum. Eng. Soc.* 12, 1985, No. 4, pp. 633-672
30 E. Rasch, Lighting Design + Application, Vol. 17/No. 9, 1987
31 H.J. Fähnrich, E. Rasch, IES-Annual Conference 1987
32 A. Dobrusskin, W. Kiele, B.Lewandowski Patent DBP 3427280, 24.7.1984
33 A. Dobrusskin, H. Leyendecker, Fernseh- und Kinotechnik 24, 1970, 11.7, pp. 222-226
34 W. Block, TechnischWissenschaftliche Abhandlungen der OSRAM-Gesellschaft Vol. 11, pp. 189-197
35 Lewandowski, TechnischWissenschaftliche Abhandlungen der OSRAM-Gesellschaft Vol. 12, 1986, p. 83
36 I.A. Kügler, *BKSIS Journal*, 1980, p. 60
37 M. Gassmann, D.L. Russell, 127th SMPTE Conf., LA. 1985
38 A. Dobrusskin, H. Leyendecker, G. Schmid, Jnt. Kongr. Reprogr. Jnf. 4th (Fachref.) 9-14, Chem. Abstr. 84, 1975
39 Philips, R., UV-Radiation, Academic Press, London, New York 1983
40 A.Dobrusskin, J. Heider, J. vom Scheidt, J. Arlt Patent DD 290 505 A5, 19.12.1989
41 Y.S. Touloukian, Thermophysical Properties of Matter, Vols. 2,8,12,13 1975
42 C. Wijenberg, Determination of the thermal conductivity of polycrystalline alumina..., Properties of Ceramics, Euro-Ceramics, Elsevier, 1989
43 J. de Groot, J. van Vliet, The HPS Lamp 240, Philips Technical Library Kluwer Technische Boeken BV, Deventer, 1986
44 S. Carleton, P. Seinen, J. Stoffels IES Conference 1996
45 S. Carleton, P. Seinen, J. Stoffels IES Conference 1996
46 Lighting Futures, Lighting Research Center NY, May/June 1995 p 1-6
47 G. Dera, E. Fischer, H. Mönch, Pys. Bl. 54(1998) Nr. 9
48 Kramer et al. *Journ. of the Electrochem. Soc.* 124, 1977, 8, p. 348
49 Yoshizawa et al. EP 81300965/359898 A 1, 16.9.81
50 Shoda, Kodama, EP 82301771/63441 A 1, 27.13.82

Chapter 9

The High Pressure Sodium Lamp

by Raymond Kane and Daniel Larson

Looking Back

The high pressure mercury discharge was recognized as a possible source of light as early as 1906 by Kuch and Retachinsky, who made studies of the discharge. Shortly afterward, attempts were made to improve the color and efficacy by the addition of other metals but met with little success. By 1930, several advances had been made, including an electrode containing thorium, a better seal of the lead wires to the glass, the addition of a fill gas to aid starting, and the addition of a fixed amount of mercury to control the voltage, and all were combined to produce the first practical lamp.

The low pressure sodium discharge lamp was known as an efficient light source as far back as 1920. Although high in luminous efficacy, it did not gain wide acceptance because of its almost monochromatic yellow color, poor color rendition and large size.

In the early 1950s, improved seals, emission materials and phosphors were developed for the high pressure mercury vapor lamp, which greatly increased lamp life, improved color and provided better lumen maintenance. It did not take long for the high pressure mercury vapor lamp to become widely accepted as a source of lighting for many outdoor applications and for high bay industrial indoor applications.

In the later part of the 1950s, General Electric resumed its work on the addition of metal halides for the mercury vapor lamp, but this time with much success. In 1961, G.H. Reiling was granted patent 3,324,421 for this achievement, and the race was on with all of the larger lamp companies diligently searching to find the best additives and proper operating conditions to make a new, highly efficient, white light source.

During that same time period, the search for an envelope that would support a high pressure sodium vapor discharge was also coming to fruition. It was known that the monochromatic color of the low pressure sodium lamp could be enhanced by raising its vapor pressure. To do this, the development of an arc tube material was required that would operate at the necessary high temperature, have a high visible light transmission and tolerate the corrosive action of the sodium vapor. Finally, about 1957, R. Coble, another GE scientist, developed a polycrystalline alumina (PCA) material.[1]

It met the prerequisites of having a light transmission of 93%, a high melting point of 2040°C, and a low reactivity to sodium vapor up to a temperature of 1300°C. GE called this material "Lucalux," and by 1958, GE's Kurt Schmidt was using it in his laboratory to make lamps filled with various alkali metals.[2,3,4]

Eight years were spent in the development of the lamp and necessary production techniques before the first lamp arrived in the marketplace. It was a 400W lamp of golden white color and an efficacy greater than 100 lumens per watt.[5]

One of the early versions of the high pressure sodium (HPS) lamp is shown in Figure 9-1.

Lucalux tubes for lamps became available in 1962, and shortly afterward, similar polycrystalline alumina tubing became available from other manufacturers. This material could not be heated and worked like quartz to make a completed arc tube with electrical feed-throughs. New methods of sealing the ends of the tube, attaching an electrical connection to the electrodes and providing a means to exhaust and fill the arc tube all needed to be developed.

The interface between the PCA and the metal connection to the electrodes also had to withstand the corrosive effects of the alkali metal at high temperatures. It also had to withstand the stress of thermal cycling as the temperature of the lamp ends increased from room temperature to 750-800°C, then subsequently cooled down when turned off, repeating this hundreds of times during its life.

Fundamentals of the Sodium Vapor Discharge

The discharge in the high pressure sodium lamp operates in an atmosphere of sodium, mercury, and a rare gas, usually xenon. The polycrystalline alumina envelope can operate as high as 1300°C with very little reaction with the sodium in the discharge.

The luminous efficacy, color, voltage and lamp life all vary with the vapor pressure of the sodium and mercury, which, in turn, are dependent upon the temperature of the cold spot in the arc tube where the sodium and mercury condense. They are also a function of the ratio of the sodium to mercury and the quantity that is in the arc tube. The luminous efficacy is also dependent upon the fill gas composition and its pressure.

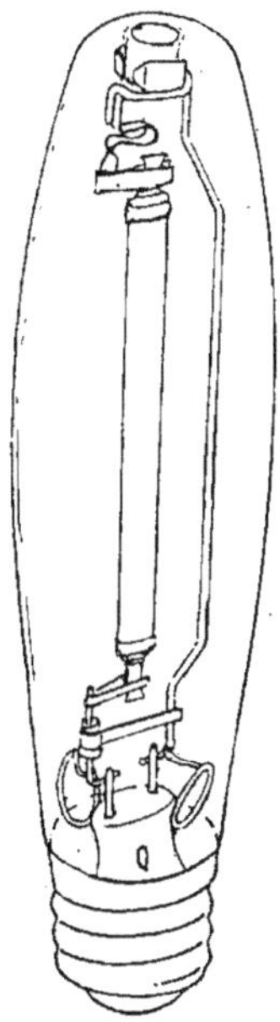

Figure 9-1. One of the earliest designs of a 400W HPS lamp.

The High Pressure Sodium Lamp Radiation Spectrum

The sodium radiation at low pressure consists of two narrow resonance lines at 589.0 and 589.6 nanometers. As the sodium pressure increases, the sodium resonance lines broaden, and by the time the pressure has reached 150 torr, they have broadened into wide bands extending tens of nanometers toward the blue and red regions of the spectrum. This can be seen in Figure 9-2. The source color looks yellowish-white.

The radiation at the resonance lines and the wavelengths adjacent to them is now strongly absorbed, producing a gap in the spectrum that extends for about 15 nanometers. This gap is the result of the absorption of radiation that is generated in the hot core by the cooler sodium vapor near the wall.

While the sodium vapor pressure increases, the luminous efficacy also increases until it reaches about 100 lumens per watt (LPW). There is little change in efficacy between 100 and 200 torr, but beyond 200 torr, it begins to decrease. The decrease is associated with an increase in the color rendering index (CRI). In an operating lamp, having an arc tube of 7.5

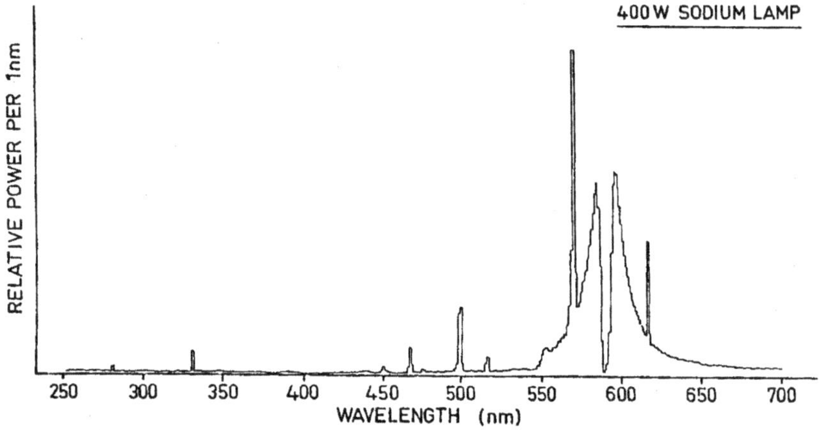

Figure 9-2. A 400W HPS spectral distribution curve showing the spread of the D-lines into the red wing and the blue wings of the spectrum.

mm inside diameter and an electrode spacing of 70 mm at a pressure of 150 torr, the lamp would normally be operating at 400W, 70V and 7.0A.

Fill Gas Studies

In his early investigations of the high pressure sodium discharge, K. Schmidt found that the different rare gases that were used as fill gases for starting, yielded different luminous efficacies in the lamps, and that the efficacy increased as the atomic weight of the gas increased. Neon produced the lowest efficacy, while xenon had the highest. He also found that with xenon, the efficacy could be further enhanced by increasing its pressure up to 200 torr, while maintaining an optimum sodium pressure. In this manner, he was able to reach 145 LPW, with a lamp of 6 mm bore, containing sodium, operating at 12A and filled with xenon at 210 torr. For this maximum efficacy, the sodium pressure had to be reduced to 75 torr and the color was shifted further toward the yellow.

The increase in efficacy results from the reduction in heat loss by conduction from the hot core to the arc tube wall. The thermal conductivity of xenon is the lowest, by far, of the rare gases and is only half of that of sodium. The addition of xenon to the sodium results in the reduced thermal conductivity of the combination. The overall effect was large because the thermal losses account for 40% or more of the power going into to the discharge's positive column.[6]

The addition of xenon introduces a band of radiation at 561 nanometers, which is near the peak of the luminosity curve. At a pressure of 200 torr, this band contributes significantly to the lumen output and improves the efficacy. The higher xenon pressure does not increase the voltage appreciably, but it does make the lamp much more difficult to start.

Mercury, whose thermal conductivity is only slightly higher than that of xenon, has a very low vapor pressure at room temperature, so its addition to the 20 torr of xenon will not affect starting.

At the lamp's operating temperature, the mercury pressure will be hundreds of torr, so it will reduce the thermal conductivity just as the xenon does. The mercury also has a pronounced effect in increasing the lamp voltage. The addition of mercury was found to increase the efficacy but only half as much as was obtained by increasing the xenon pressure to 200 torr.

The Sodium Amalgam Effect

The presence of mercury provides more radiation in the red wing, shifting the color towards white, but with some reduction in spectral efficiency. At the same time, the total wattage radiated in the visible region increases and results in an overall increase in efficacy.

Mercury and sodium, when mixed together, form an amalgam in which the vapor pressure of each is less than it would be as separated elements. The loss of sodium by its reaction with other lamp components results in an increase in the mercury pressure, which in turn, increases the lamp voltage. A large excess of sodium-mercury amalgam is added to the lamp to minimize the effect of sodium loss on the amalgam ratio and, therefore, the voltage.

Changing from pure sodium to a 21% by weight sodium amalgam almost doubles the discharge voltage if the sodium pressure is unchanged. This results in lower losses at the electrodes and makes possible the use of a smaller, lower-loss ballast.

On currently used ballasts, as the sodium is depleted, the lamp voltage, wattage and arc tube temperature increase, further accelerating various sodium reactions. Adverse reactions occur with the alumina tube, the end-seal frit and the electrode emission material.

Over the years, improvements in the PCA material, the monolithic tube design, the end sealing material, the emission material and the increased xenon pressure, have greatly reduced the reactions which result in the loss of sodium. These, combined with operation at slightly lower wall temperature and seal temperature, have resulted in lamps in which the amount of sodium lost by reactions became very small.

Consequently, the amount of mercury and sodium in the amalgam could be reduced to the point where reasonable life times could be achieved without the lamp voltage increasing to the point at which the lamp will extinguish. This was achieved by reducing the amount of sodium to a small fraction of that used in conventional lamps and the mercury to an amount, which even if completely vaporized, would not cause the lamp to extinguish.[7]

The PCA Tube

In the pursuit of a viable arc tube for the high pressure sodium lamp, Cobel succeeded by adding about 0.2% of magnesium oxide to an aluminum oxide powder of very high purity (99.9%). Any further growth of the fine grains of aluminum oxide, which were of the size of 10μ to 30μ, was inhibited by the presence of the magnesium oxide, thereby allowing the alumina to be sintered densely enough to provide adequate translucency with a light transmission that is a remarkable 93%. The finished tubing is so hard that it must be cut with a diamond wheel. Unfortunately, it has a sharply defined melting point, thus providing relatively no plastic working range. The open ends of these cylindrical polycrystalline alumina (PCA) tubes, therefore, must be sealed with a separate suitable cap or disk.

Disk and Sealant

The disk and its sealant must be resistant to the corrosive effects of the sodium at high temperatures and be able to withstand the thermal cycling.

The early lamps made by Schmidt in 1958 used a nickel-chrome-iron alloy disk which was bonded to the PCA tube with a thin wafer of titanium. A second titanium wafer and a PCA ring were then bonded to the outer surface of the alloy disk. Many other unique variations of laminated metal disks were tried, but even though their coefficients of expansion were cleverly formulated to provide a near-match to the PCA tube, they were ultimately abandoned. Various end closure designs are shown in Figure 9-3.

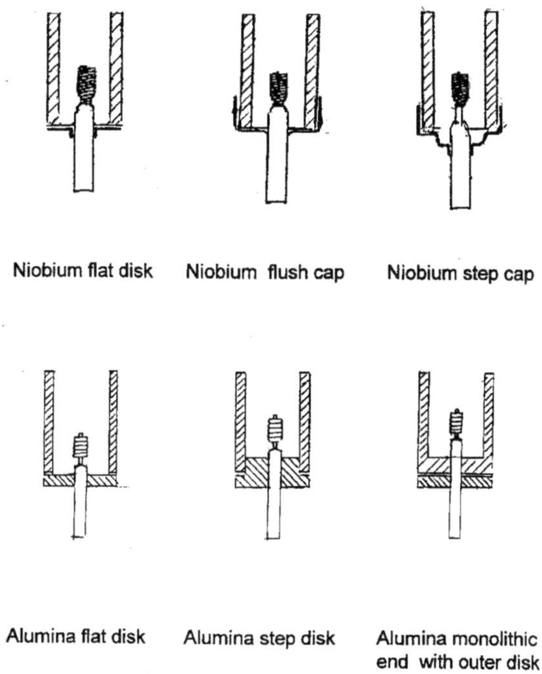

Figure 9-3. Various arc tubes, some of early design, showing differences in their end assemblies.

The metal disk concept was superseded by a thimble-shaped cap of niobium which was sealed to the PCA with a calcium aluminate glass having a melting point near 1400°C. The cap had to be a close mechanical fit to the PCA tubing so that the glass frit would flow between the two by capillary action. This required tight control of the end shape and size of the tubing and of the cap. Such a construction provided a longer path through the sealed region which inhibited the escape of the corrosive sodium vapor, and provided some resiliency in the cap to reduce the strain on the seal. Control of the seal's melting process was critical so the bond between the frit and the cap was not consistently reliable. To improve adherence, the bonding surface of the caps was sandblasted and sometimes coated with thin layers of other refractory metals. Some small improvements were achieved with a thin paste of niobium, silicon or tungsten oxide powder, but none of these significantly reduced the eventual sodium loss.

The next end closures tried were disks and plugs made of PCA. New formulations of the glass frit were developed to improve the bond. Some disks were formed in a step design in which the collar fits into the PCA tube and the lip extends over the end of the tube. All of the end plugs were made with a centrally located hole through which an exhaust tube was then inserted.

A further improvement came with the *monolithic* tube, in which the end plugs became an integral part of the tube. This was achieved by attaching a pre-fired disk to each end of a pre-fired PCA tube. A subsequent firing produced a homogeneous union between the disks and the tube wall.

In a later modification, an additional disk was cemented over the monolithic end to increase the seal length along the niobium tube wall to make it more difficult for the sodium vapor to work its way through the sealant.

The PCA Tube Manufacturing Process

PCA tubes are manufactured under very rigorous clean room conditions. The flow chart in Figure 9-4 shows how PCA tubes are typically made.

A fine alumina powder, about 0.5 micron in size, is first made into a slurry by adding distilled water, an organic binder, and magnesium oxide. The slurry is next passed through a spray dryer to cluster the alumina particles. These agglomerates are then sieved to obtain the proper size distribution. Those of proper size are then sent to an automatic weighing machine which measures the proper amount of powder required to make an individual tube. Next, the powder is either isostatically pressed around a stainless steel mandrel or made into a paste and forced through an extrusion die, to be formed into the proper cylindrical shape. At this point, the tubes are in their green state and are extremely fragile. However, they are porous and are easily cut to length, with an allowance made for the reduction in size that will come with subsequent heat treatments.

The tubes are then pre-sintered in air at about 1100°C for approximately two hours to remove the binder and to improve their strength for further handling. Next, they are sintered in hydrogen for six hours at about 1800°C (alumina's melting point is 2040°C). During the hydrogen sintering process, the tube will shrink in every direction to about 70% of its former size. The sintered tubes are then cut with a diamond wheel to

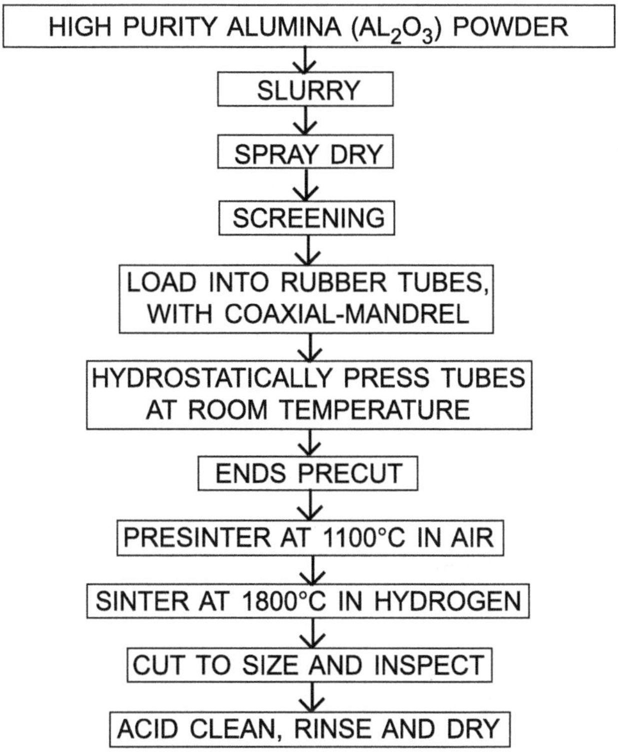

Figure 9-4. A typical flowchart showing how PCA bodies can be produced.

the desired length. The cutting process provides them with smooth ends that will resist chipping during subsequent operations.

Extensive tests and measurements are then performed on finished tubes selected from each powder lot. Tests are made on hardness, concentricity, squareness of ends, degree of porosity, uniformity in color, straightness and pertinent physical dimensions.

Improvements in the PCA Tube

Although PCA quickly gained wide acceptance in the production of HPS arc tubes, extensive engineering work continued in order to improve the tube quality. Problems such as micropores and microchips had to be eliminated, as well as non-uniformity of the tubing wall, tapering of the ends, thermal cracking, bowing and crookedness of the longer length pieces, and opaque white spots called "blebs." Perhaps most important was the necessity to obtain a tight control of the inside diameter so that every tube could be fitted with a ceramic end plug, button or exhaust tube, with a bare minimum of clearance, to make it more difficult for the sodium to escape.

Companies such as Coors and Corning of the USA, NGK of Japan and Ceraver of France did extensive work to meet the ever-increasing demands of the lamp companies for more and better material. Eventually, because of the enormous volume and relatively high cost, most of the major lamp companies set up their own PCA tubing production facilities.

It was found that the light transmission of the PCA could be increased by chemically polishing the outside surface to make it smoother. Unfortunately, the benefit derived was offset by the added cost. However, J.G.J. Peelen of N.V. Philips devised a method of hot-pressing the normally sintered tubes that reduced the grain size and pore size of the alumina, thereby making it more transparent.[8]

Sapphire Tubes

Tubes have also been made successfully from sapphire which is drawn as a single crystal from molten alumina. Initially, solid boules of sapphire were drawn and the core was removed with diamond drills. The inner and outer surfaces were then ground to their desired dimension and then polished. This was an expensive process for obtaining clear sapphire tubing.

In the mid-1960s, the Tyco Corporation developed a method of drawing crystalline sapphire tubing with well-controlled inner and outer diameters. This greatly reduced the cost and increased the lengths of the tubing that could be made.

From 1972 until 1975, a good portion of the HPS lamps made by the Westinghouse Corporation were made with sapphire tubes.

The luminous efficacy of the HPS lamps made with single crystal sapphire arc tubes is improved by about 5% over that obtained with polycrystalline alumina tubes. In addition, sapphire has no grain boundaries to allow the sodium ions to migrate through the tubing wall. It is also almost totally transparent, which made it useful for many of the original experiments that had to be done to view and evaluate crucial lamp performance characteristics. These included measuring electrode and arc discharge temperatures, making luminance profiles of the arc discharge extending from its hot core to the bulb wall, as well as investigating other plasma phenomena.

In spite of sapphire's optical superiority, serious problems existed such as end-chipping, seal cracking, and in making a viable, single crystal tube. It also had inherently higher costs that were related to the slow process of growing the tubular crystal from its molten source. Often, just the least amount of vibration during the drawing process would cause more than one crystal to form. The multiple grain boundaries were often deeply fissured, which led to leakage and cracking. All of these factors discouraged its commercial use for making sodium vapor lamps, especially as the demand for the lamps sky-rocketed.

Development of the Arc Tube

The HPS lamp arc tubes evolved in their design with changes that reduced the sodium losses through the seal, improved the purity and consistency of the sodium amalgam dosage, improved the emission material, altered the design of the electrode and optimized the arc tube dimensions.

At the IES Conference in September 1966, Louden and Matz of GE described a line of six lamps ranging in size from 125-1000W.[9] Gradually, the line expanded to other wattages, arc tube diameters, lengths, improved methods of closing the arc tube ends, and many configurations of tantalum and niobium end terminals.

The novel, and sometimes dramatic, modifications of the HPS arc tube components resulted in a steady improvement in the finished lamps.

The Sealing Material

All PCA tubing ends must be closed hermetically with disks or plugs that are bonded to the tubing wall before or when the electrodes are inserted. A refractory sealing compound, or "frit," is used to provide this bond. It must be very resistant to the highly corrosive action of the hot sodium vapors when the lamp is operating, and it must be able to withstand the severe thermal shocks which occur when the lamp is turned on or off. Its coefficient of expansion must be a good match with those components to which it is bonding.

For HPS lamps, the frit must have a melting point lower than the softening points of the alumina and niobium, but must be higher than 900°C, which is the maximum operating temperature found at the arc tube ends. Another consideration is that the frit must not allow oxygen or other contaminating materials to penetrate the arc tube, as determined by Luthra.[10]

Compounds that have been developed for use as sealing frit generally consist of a near-eutectic blend of alumina and calcia, with small amounts of silica, magnesia, barium oxide, boron oxide or other refractory oxides which are added, primarily, to reduce the grain size. The percentage of each material that is added to the sealing compound must be critically controlled to obtain the correct melting point. The actual temperature of the melting and solidification points is confirmed by using differential thermal analysis, or DTA with a typical example shown in Figure 9-5.[11,12]

Frit pieces are usually fabricated as circular rings because of their ease of handling, forming and storage. To give the rings good flowing

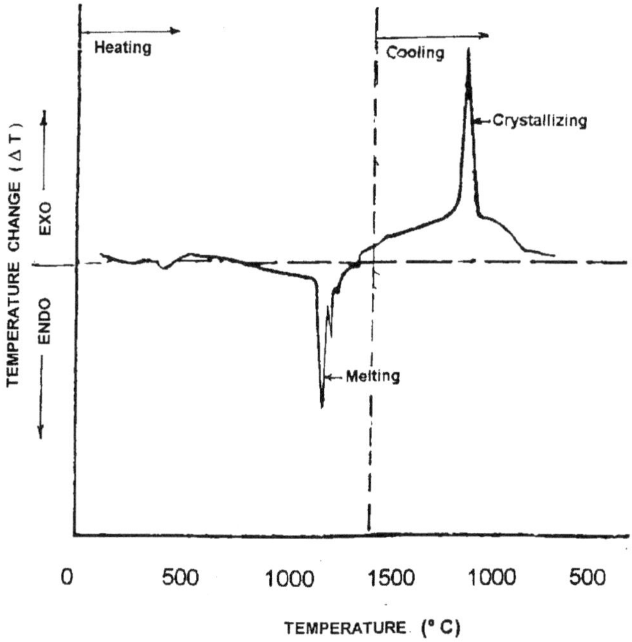

Figure 9-5. A differential thermal analysis showing the melting and solidification temperature peaks of a sealing frit mixture.

properties, integrity, and stability during manufacturing, additional ingredients must be added. Among these are usually a binder, a solvent, a deflocculent and a dispersent.

The frit mixture is first blended, then melted as a batch, solidified, and then crushed and pulverized. The fine powder is then pressed and formed into rings. Next, the finished rings are baked in an oxygen-enriched atmosphere to remove the binders and the other volatile materials.

Before frit rings became available, a slurry was used, which had to be laboriously brush-painted onto the narrow edge of the PCA tube, just prior to attaching the end plug. It was difficult to keep some of the slurry from oozing into the body of the arc tube.

The inner and outer diameters of the frit ring usually approximate those of the PCA arc tube. The ring thickness determines the volume of the frit needed to obtain a neat and tight seal.

Selecting the proper size and weight of the preformed frit ring is important. Using too much frit will result in some of it flowing inside the arc tube. This is undesirable because the frit is less resistant to chemical reaction with sodium than PCA is, so it contributes to the loss of sodium from the amalgam, thereby changing the amalgam ratio which controls the color quality and voltage.

In addition, some frits react with the niobium and become conducting. If the amalgam is condensed on the end disk, it can be electrically connected to the electrode. In this condition, the arc can strike the amalgam rather than the electrode and cause thermal shock resulting in cracking of the PCA tube.

Too little frit will leave voids in the seal which will eventually provide a leakage path for the sodium.

Many studies were made with the scanning electron microscope and other diagnostic instruments to learn how the vast number of frit crystal structures were formed, what determined their pattern of growth, and how they reacted with the other chemicals in the seal region to produce the necessary long-term adhesion required to bond the PCA tube, the end plugs and the niobium tubes. To this day, the eventual loss of the seal's integrity is still one of the predominant causes of lamp failure, although it occurs much later in life.

The Electrodes

In the early stages of development of high pressure sodium lamps, most of the work was directed toward making experimental lamps in the 400W and 1000W sizes. Test electrodes, actually made for production mercury vapor lamps were readily available. Furthermore, although the 1000W arc tube was longer than the 400W, the inner and outer diameters were the same, thereby making it possible to work with a PCA envelope of only one cross-section. Much of the development effort concerned the optimization of the other parameters, such as arc length, the quantity and ratio of sodium and mercury, the composition and pressure of the fill gas, and to establish suitable methods of manufacturing the lamps in production quantities.

Gradually, it became apparent that the HPS lamp electrodes had to have their own unique construction, primarily because of their influence on determining the cold spot temperature of the sodium-mercury amalgam.

Basically, a typical HPS arc tube electrode (Figure 9-6) consists of a high-quality tungsten rod that has been doped with 1-2% of thorium oxide, around which is wrapped a continuous wound, two-layer tungsten coil which does not contain any thoria. The addition of thorium oxide into the coil wire itself was found to provide no added benefit. The coil of two layers is purposely wound back and forth with a single continuous wire to avoid the exposure of the sharp points of the wire ends that would otherwise face the tip of the rod, where the arc is most likely to strike upon ignition. This would result in an undesirable sputtering of the tungsten onto the arc tube wall.

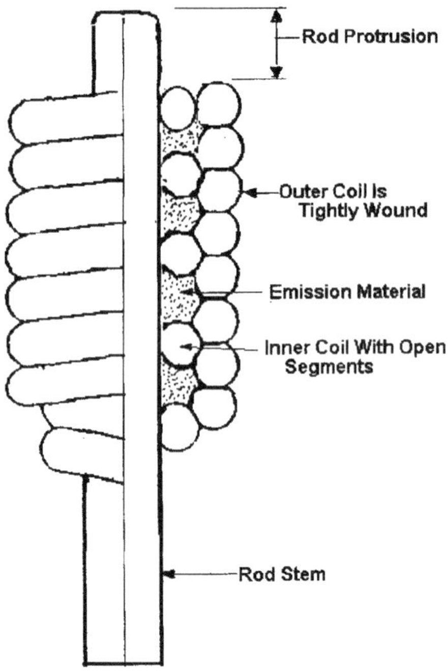

Figure 9-6. One common electrode assembly showing the relative position of the rod to the coil barrel and the inner coil spaces formed to hold the emission material.

The inner coil layer is wound with open spaces at its

center which provides pockets to hold the emission material. The outer winding is wound very tightly in order to retain the emission material and to protect it from bombardment by the arc discharge. This so-called back-wound coil, is made from lamp quality, "AKS-doped" tungsten, selected because of its formability and high temperature creep resistance. (The term "AKS" is the chemical acronym for aluminum, potassium and silicon, whose salt compounds are added in minute quantities to achieve the desired configuration of the tungsten grains.)

The distance that the rod protrudes beyond the forward edge of the coil must be carefully controlled in order to obtain the correct tip temperature.

The actual parameters of the rod and coil must be established by making many compromises. One major consideration is the fact that the lamps are operated on alternating current circuits, which means that the electrodes must alternate as cathode and anode. When the electrode is acting as cathode, it must supply the required electrons to the arc. These electrons are emitted from the front of the electrode, which is heated by positive ions during the cathode half-cycle, and by electrons bombarding it during the anode half-cycle.

The emission material embedded in the spaces between the coils and the center rod must be hot enough to supply barium to the front of the electrode. The discharge terminates on the face of the rod. The anode heating helps in obtaining the temperature required for the electrode to operate effectively as a cathode.

During life, the "hot spot," which is the point on the electrode where electron emission occurs, gradually moves down the coil, seeking out the depleting supply of emission material. As this occurs, the electrode power consumption increases, as does the length of the arc discharge, with both factors contributing to the voltage rise of the lamp.

The electrodes must be designed to function properly during all three of the vastly different operating conditions which develop in the arc tube. These occur during starting where there is a pulse breakdown, during a warm-up or "glow" period of high voltage and low current, and then finally at a stable or full-arc discharge state.

The transition to a stable low-voltage, high-current condition with good electrodes can occur in less than a second. Toward the end of the lamp's useful life, it may take up to a minute, during which time sputtering and evaporation of the tungsten electrodes will be going on.

Provision must also be made to achieve good performance and

long life in spite of wide variations in operating currents that result from normal lamp and ballast manufacturing tolerances.

The power consumed by the electrodes is of no practical value for the production of visible light, so it is desirable to minimize this loss, that averages about 24 watts for a 400W lamp. Studies have shown that the electrode loss can be kept low through the use of good emission material and by the proper selection of rod diameter, rod protrusion, coil size and overall electrode length.[13,14]

The Emission Material

In the HPS lamp, as with most other discharge lamps, an emission material is added to the tungsten electrode to reduce its work function. When this occurs, the cathode potential drops, causing a reduction in the velocity at which ions strike the electrode, which lowers its operating temperature. As a result, the electrode's evaporation and sputtering losses are substantially diminished.

Besides lowering the work function by its efficient reaction with the tungsten substrate, Bhalla reports that the emission material must also exhibit the following traits:[15]

- Its melting point must be high enough to prevent chemical changes from taking place during the arc tube sealing process.
- During arc tube manufacture, it must be very stable at room temperature, and not react with moisture or other contaminants while it is exposed to air.
- It must have a low evaporation rate at the electrode's operating temperature.
- It must not react with the metallic vapors in the arc stream (i.e., mercury and sodium) at the arc tube's operating temperature.
- If sputtered, evaporated or otherwise injected into the arc stream, it must not react with the PCA body or the end-sealing materials.

Originally, the HPS emission material was that used in high pressure mercury vapor lamps. It was a blend of the carbonates of barium, calcium and strontium, to which some thorium oxide was added. This mixture was then heated at a high temperature to convert the alkali carbonates to oxides.

Subsequent investigation showed that in order to meet the rigorous criteria outlined above, differently formulated emission materials were required. Furthermore, there was a compelling interest to eliminate the potentially harmful, radioactive thoria.

Several new emission materials have been developed for the HPS lamp. Two that have been widely used in this country are tribarium yttrium tungstate and dibarium calcium tungstate. Outside of the United States, other refractory compounds, such as beryllium oxide, are sometimes combined with a tungsten powder to form an emission mixture.[16]

More recently, however, we have seen the introduction of the tungsten-free emitters of various barium strontium yttrate compounds which are used with the unsaturated lamp designs. These show promise in providing lamps of longer life, improved lumen maintenance and better voltage stability.[17]

Emission Material Preparation

Selecting the best emission material composition does not, in itself, ensure successful electrode performance. Of utmost importance is its proper preparation. Stoichiometric quantities of the basic oxides must first be uniformly blended, then ball-milled in an alcohol or similar suspension until the correct powder size is obtained. The resultant slurry is air-dried and then fired at a high enough temperature to achieve the desired chemical composition. The finished compound is then checked by X-ray diffraction to ensure that no extraneous phases or compounds were produced.

The emission material in powder form must then be mixed with an appropriate vehicle, sometimes something as simple as ethanol or perhaps, as a lacquer consisting of a small amount of nitrocellulose in a solution of butyl acetate. The final slurry is thoroughly milled for several hours to obtain the right consistency. It is then carefully measured for viscosity and density.

It was found that maximum weight, or "packing," can best be achieved by applying the emission material to the coil in an evacuated enclosure. For example, a large number of clean, tungsten electrodes are placed on a small shelf inside a stainless steel vessel into which was previously poured an appropriate amount of emission material. After the vessel is closed, it is then evacuated to remove the air between the interstices of the coil turns.

The vessel is then overturned, causing the electrodes to fall into the slurry, whereupon all voids in the coils are immediately and completely filled with the emission material. The vessel is then opened and the coated electrodes are carefully removed and allowed to dry in air. They are then baked in hydrogen at 1600°C for about 30 minutes.

The electrodes are removed from the furnace, allowed to cool, then either brushed or tumbled in a container of fine tungsten powder so as to remove the emission material adhering to the outer surface of the coil. Failure to remove the emission material deposited on the coil's outer surface will result in its being sputtered onto the inner wall of the arc tube when the lamp is lighted. Over time, this will cause the arc tube ends to blacken, resulting in a significant increase in the lamp's operating voltage.[18]

The final amount of the emission material embedded within the electrode must be monitored on a daily basis to ensure that it meets the weight requirement. Coils are cross-sectioned regularly to examine the emission material within the windings for proper color, amount and extent of bulk reduction. The quality and performance of the emission material is constantly monitored by testing finished lamps for ease of starting, voltage rise, onset of end blackening and duration of life.

The Electrode Support

Over the years, many techniques have been tried to support the HPS arc tube electrodes concentrically within the PCA tube, and also provide a means through the opening at the tube ends to complete the electrical circuit.

Originally, stainless steel and a while later, tantalum tubes were used, to provide the electrical feed-through, and to provide the means for evacuating the arc tube.[19]

Then it was found that niobium was a better material because its coefficient of expansion is a very good match to the alumina envelope. It is also very ductile and can be easily fabricated into a wire, cap or tube form.[20]

Niobium is a good getter for hydrogen and also acts as a porous membrane because it allows the passage of hydrogen from a high concentration environment into one which is lower such as the lamp's outer jacket. There the hydrogen is effectively gettered by the layer of barium getter that is deposited on the jacket's outer wall near the base. Using niobium tubes instead of solid niobium wires as electrode supports and feed-throughs results in much quicker removal of hydrogen from the arc tube.

Positioning the Niobium Tube

Novel methods have been devised of forming a "stop" or fixed point on the niobium tube or wire feed-through in order to set the critical distance between the electrode tip to the inside face of the PCA tube end. A variation no greater than 0.25 mm must be maintained in order to control the temperature of the amalgam that is condensed behind the electrode. This is where the cold spot is located, and its temperature determines the vapor pressure of sodium and mercury in the amalgam.

In previous designs, when niobium end disks or caps were used, the support distance to the arc tube ends could be carefully controlled by using welding fixtures that held all parts precisely in place as they are being welded. However, because of their high cost and greater difficulty in obtaining an enduring seal, the use of niobium end caps was eventually abandoned. With the introduction of the alumina end cap and the monolithic tube, new methods were required to control the position of the niobium tube.

To fix the support distance when a solid piece of niobium terminal wire protrudes from the alumina disk, all that is needed is a short cross-piece of niobium wire. However, when the terminal is a round niobium tube, a short cross-piece of wire cannot be attached to it well-enough to accurately position the tip of the electrode. Therefore, although various wires and rings have been tried, it was found advantageous to alter the shape of the niobium tube itself, to obtain the precise support distance. This can be done by forming a doughnut bulge around the tube, or by scraping concentric burrs into the niobium tube's outer surface (see Figure 9-7).

Electrode Axial Alignment

Good axial alignment of the electrode with the niobium tubulation is absolutely necessary in order to maintain a uniform temperature on the adjacent bulb wall. The narrower the bore of the PCA tubing, the more critical it is to control the alignment. Axial mounting of the electrode is usually achieved by slipping its central rod into a hole made at the end of the closed niobium tube. Sometimes, the electrode rod is butt-welded to a short niobium terminal

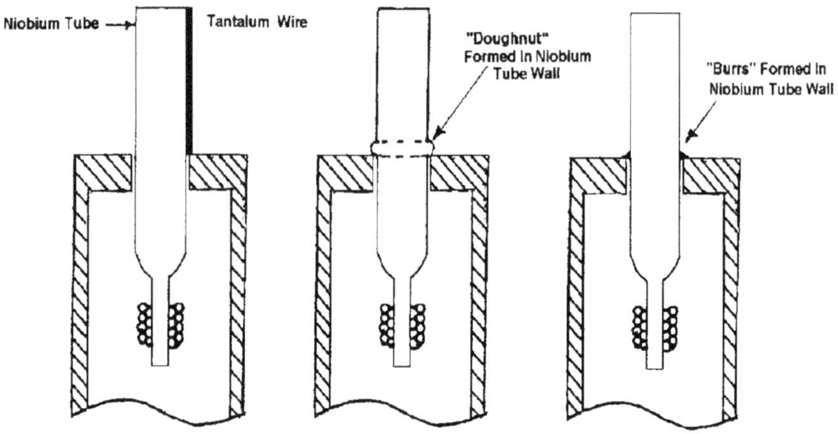

Figure 9-7. Various methods used to keep the niobium tube from falling through the PCA end disk.

wire, which then is passed through the hole in the ceramic end disk.

However, a niobium wire is not as rigid as a niobium tube and is more likely to bend during processing or when in service. Nevertheless, arc tubes of very narrow bore must use the wire terminal because the niobium tube becomes too large to fit into the hole of small alumina end disks.

One practical method of bonding the electrode to the niobium tube is to braze them together using a piece of titanium wire. After the electrode is snugly inserted into the niobrium tube and carefully oriented, the titanium piece is dropped in. The titanium piece is melted at 1800°C in an inert gas atmosphere. When finished, the entire assembly becomes firmly and hermetically joined together.

To prevent problems from occurring during lamp operation, the titanium braze must completely surround the electrode rod at its junction with the tube to make a good hermetic seal and to make a good mechanical bond to support the electrode.

The Amalgam

The HPS lamp arc tube is dosed with a small quantity of sodium and mercury; the sodium to produce the highly efficient light, and the mercury to increase the potential gradient of the arc. The two metals are combined in the ratio that is required to achieve the desired voltage and color rendition at the highest efficacy. Because some of the sodium is gradually "cleaned-up" or chemically combined as the lamp ages, an excess amount of amalgam is usually dosed into the arc tube.

Method of Dosing

After a long search to find a convenient method to insert the highly reactive sodium into a mercury filled arc tube, a method was devised to melt the sodium and mercury together in a stainless steel chamber. After mixing them thoroughly, the liquid amalgam is passed through a calibrated capillary tube into a chilled atmosphere. The amalgam leaves the tube in drops of the desired size and solid spheres of the correct weight and composition are formed as they cool.

As an alternative, the molten amalgam could be passed through the capillary tube and dropped directly in the arc tube. Unfortunately, it was found that prolonged containment of the amalgam within its dosing chamber would result in a gradual change in its composition due to the settling of the heavier mercury from the much lighter sodium.

When a substantially high percentage of sodium is used in some amalgam formulations, on the order of 25% or more, the amalgam will become sticky at room temperature. Therefore, to keep the spherical pellets from sticking together, it is necessary to work in a refrigerated atmosphere to make certain that the pellets do not stick together and that the correct quantity is dosed into the arc tube.

Amalgam Characteristics

For most general lighting lamps, the optimum luminous efficacy is obtained with a weight ratio of about 21-25% sodium to mercury.

At 21% weight, the mole fraction of sodium present in the amalgam is 0.70. This ratio of sodium will produce the desired vapor pressures of 100 torr for sodium and 700 torr for mercury, when the cold spot temperature of the amalgam is 750°C.[21] Schmidt shows how, at 700°C, the vapor pressure of mercury and sodium vary as the amalgam ratio changes (see Figure 9-8).[22]

The sodium and mercury vapor pressures are determined by their ratio in the amalgam and the amalgam temperature. The effects due to varying the pressures can be seen by varying the cold spot temperature as Denbigh shows in Figure 9-9.[23]

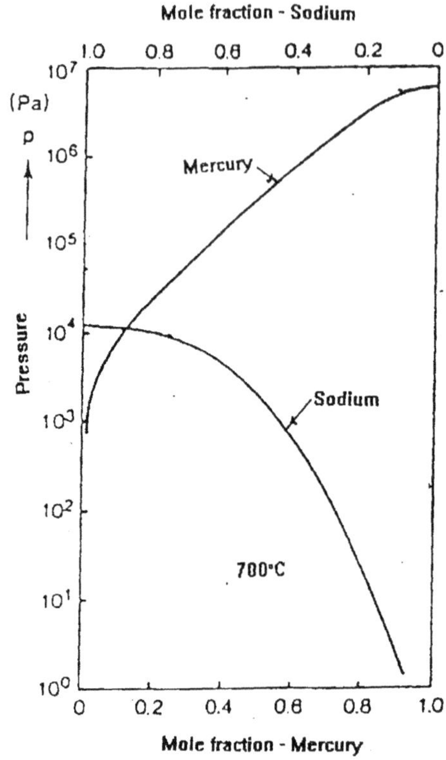

Figure 9-8. Curves showing the change in sodium vapor and mercury vapor pressures as the amalgam ratio is changed (from K. Schmidt, U.S. Patent 3,384,798, 1968).

With the lamp characteristics being determined by the vapor pressures of the mercury and sodium, it would be useful if these pressures could be determined from the spectra of completed lamps.

Knowing that the sodium resonant lines were broadened and strongly absorbed at, and adjacent to, their central wavelengths by the sodium vapor, it followed that the wavelength difference between the two peaks, a value referred to as "delta lambda," or simply, $\Delta\lambda$, gave a good indication of the relative change in sodium pressure (see Figure 9-10, top). Later, it was learned that the region in the "blue wing" peak of the doublet was not influenced much by the presence of the mercury atoms. Therefore, by measuring the distance $\Delta\lambda_B$ between the peak of the blue wing and the mid-point between the unbroadened resonance lines at 589.3 nanometers, a more accurate measure of the sodium vapor pressure could be made.

Likewise, the distance $\Delta\lambda_R$ helps to determine the mercury pressure in the lamp (Figure 9-10, bottom).

During lamp life, the amalgam composition can change due to sodium clean-up, and the temperature of the cool spot controlling the pressures of the sodium and mercury can also change. Measurements of $\Delta\lambda_B$ and $\Delta\lambda_R$ can be made to indicate the specific amount of sodium loss and temperature increase that are occurring during ageing of the lamp.[24]

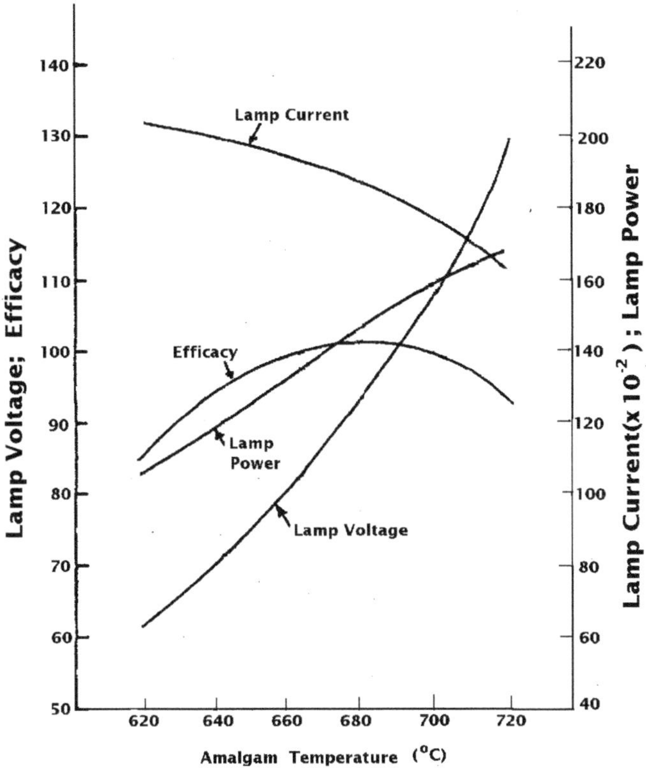

Figure 9-9. Curves showing how various lamp characteristics are affected by a change in the cold spot temperature (from Denbigh, 1978). Lamp is rated at 150W, 100V. Sodium mole fraction is 0.688. Courtesy of Lighting Research & Technology.

The Arc Tube Assembly

Although the fabrication of the arc tube is straight-forward, experience has shown that it is vital to seal both ends of the arc tube in a clean room type glove box. This insures that the amalgam and electrodes are not exposed to moisture, oxygen, particulates and other impurities that could end up in the finished arc tube.

The glove box is usually filled with high purity argon gas at a pressure of one atmosphere to maintain equilibrium with the outside atmosphere. The argon gas is constantly circulated and cleansed by

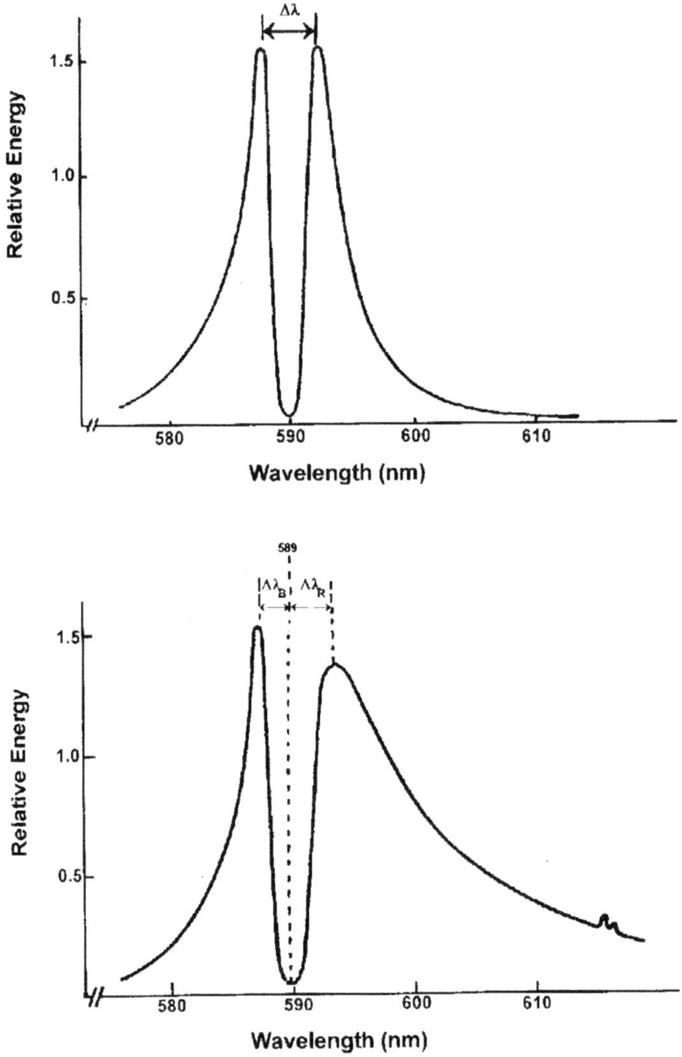

Figure 9-10. TOP. The distance between the two peaks of the sodium resonance lines, referred to as $\Delta\lambda$, as measured in a lamp containing only sodium. BOTTOM. The distance from the sodium resonance line at 589 nm to the peak of the blue wing, $\Delta\lambda_B$, and also to the peak of the red wing, $\Delta\lambda_R$, in a lamp containing both sodium and mercury, and used to estimate the sodium and mercury vapor pressures within the arc tube (from Jacobs et al, 1978). Courtesy of the Illuminating Engineering Society of North America.

passing it through a molecular sieve dryer which will usually contain an oxygen scavenger such as phosphorus pentoxide.

All of the materials enter the glove box through an antechamber which is then evacuated and filled with argon. The materials are then passed through a port into the main glove box.

The Sealing Process

The frit ring and ceramic disk, together with a niobium post-to-electrode sub-assembly, are placed on top of one end of a vertically positioned PCA tube (see Figure 9-11). The entire assembly is gently placed into a high temperature vacuum furnace connected to the glove box. The temperature is raised at a pre-determined rate to approximately 1450°C, at which point the frit ring will melt and run between the mating surfaces of the PCA and niobium. The cooling is generally done in stages, and at different rates, so that the correct crystalline structures are formed at the PCA-to-disk-to-niobium tube interfaces.

For efficiency and cost savings, many tubes are sealed at the same time. They are usually loaded into a water-cooled molybdenum holder which is raised into a furnace that surrounds the holder and provides a hermetic seal for the subsequent evacuation of the chamber. The furnace heating element can be a block of carbon formed to have the same proximity to each of the arc tubes so that all of the frit rings will melt at the same time.

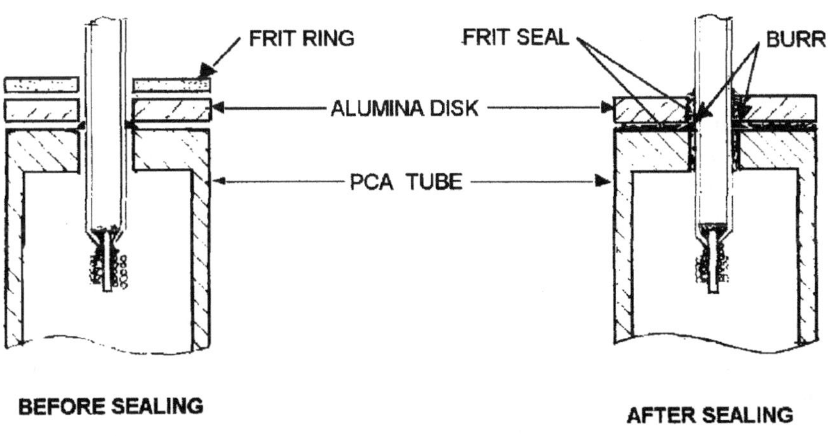

Figure 9-11. The components of the end assembly of the HPS arc tube before (left) and after (right) they are sealed together.

Before the frit seal heating cycle is finished, a point is reached when an inert gas, such as argon, is admitted into the sealing chamber to speed up the cooling. When the tubes are cool enough to be handled, they are removed from the furnace.

This first seal is then examined for color, frit flow quality, smoothness and uniformity. Arc tubes with defective or poor-looking seals are removed from the lot and discarded.

After inspection of the first seal, the arc tube is allowed to cool to room temperature while still in the glove box. Sometimes a small refrigeration unit must be installed inside of the glove box to get the temperature low enough to where it is comfortable for the operator to work with her gloved hands and to keep the sodium amalgam from softening.

Next, the correct amount of the amalgam is dosed into the arc tube which is surrounded by a water jacket at its lower end to keep the amalgam from vaporizing during the subsequent sealing operation. The second set of end parts is then carefully placed upon the open end of the PCA tube, making sure to maintain a good concentricity with the PCA tube.

The tubes are then loaded into the furnace to make the second end seal, but with this operation, xenon gas is forced into the evacuated arc tube just before the frit ring melts. This precise method of manufacture must be timed with a programmable controller.

The finished arc tubes are removed from the glove box, by way of its antechamber, then the seals are more carefully examined under a magnifying glass for imperfections. They are then placed in an orderly fashion in a shallow tray and excited with a high frequency coil. Defective tubes are easily recognized because the color of the excited gas in the arc tube is different from the pale blue that is seen with pure xenon. Arc tubes with physical defects, such as crooked ends or loose electrodes are also culled out.

All finished arc tubes must be handled with lint-free cotton gloves because body salts and oils will attack and discolor the tubes when they are operating at temperatures reaching 1200°C.

Earlier processing techniques consisted of slow, tedious methods of assembly, dosing and sealing parts in unpurified factory air, then exhausting and gas-filling the sealed arc tube through a niobium exhaust tube. The niobium tube had to be pinched flat very tightly, then seam-welded to obtain a strong, hermetic seal. It was necessary to store the arc tubes for up to two days, then recheck them with the high frequency coil to make sure that leaks had not developed.

Lamp Starting Aids

With its small arc tube diameter and 20 torr xenon as a starting gas, the HPS lamp will not start on mercury or metal halide ballasts. It requires a high-voltage pulse to create the initial breakdown of the xenon. The pulse for the 400W lamp, for example, is specified (by ANSI) as being a minimum of 2250V and one microsecond in duration.

With the continuing desire to improve lamp performance, a greater number of lamps are being made with an increase in xenon fill pressure from 20 torr up to as much as 200 torr, where an additional 10% increase in lumen efficacy can be achieved.

However, the lamps then become hard to start, even with the presence of the starting pulse. This can be overcome by placing a starting aid of tantalum or tungsten wire against or around the arc tube, and connecting one end of the wire to one of the electrodes. Thus, the breakdown path between the electrodes is substantially reduced capacitively, and allows the lamp to start on a conventional HPS lamp ballast.

When the lamp starts, the wire must be removed from the electrical circuit and/or the hot PCA tube by using a bi-metal switch, a diode, capacitor or similar means, so as to prevent the electrolysis of sodium through the PCA wall. Many wire configurations can be used, such as the one shown in Figure 9-12.

In some instances, such as where sodium lamps are desired to retrofit into existing mercury lamp fixtures, it is not practical to install a separate starting device. Instead, a neon-argon gas Penning mixture, generally 99.7% neon and 0.3% argon, is used as the starting gas in place of xenon.[25] Such a lamp design results in as much as a 20% loss in lamp efficacy, but its light output is still greater than that of the mercury lamp it replaces.

More recent retrofit lamps, such as the Philips "Retrolux" line, are being made with an ignitor mounted within the outer bulb and do not use the Penning mixture gas fill.

The Outer Bulb Assembly

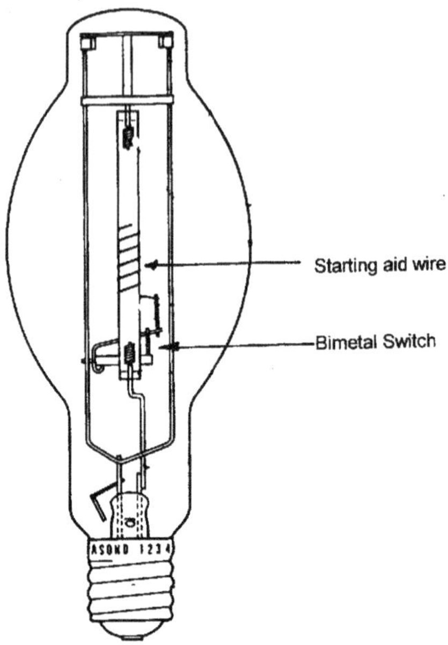

Figure 9-12. A tungsten wire used as a starting aid; it is removed electrically from the arc tube after the lamp starts. Courtesy of Philips Lighting Company.

After a careful inspection for overall quality, uniformity and appearance, the finished HPS arc tube is mounted to a nickel or iron frame which must be sturdy enough to withstand the vibrations and shocks often encountered in lighting fixtures, which, if severe enough, may cause the arc to extinguish. One end of the arc tube is generally secured to the lamp stem within the bulb, but the opposite end must be allowed to move axially, to avoid being stressed by expansion when the lamp is turned on.

The frame must support the arc tube in a position that holds it concentric within the outer bulb and along the optical axis of the lighting fixture.

In earlier designs, the frame wire adjacent to the arc tube was covered with a quartz sleeve to retard the emission of electrons from the barium-coated nickel-iron frame, generated by the radiation from the arc tube (the barium came from the flashed getter). The electrons from the frame wire collected on the arc tube, giving it a negative charge, causing the positive sodium ions to migrate through the PCA wall.[26] A mount construction made with a photo-electron collector is shown in Figure 9-13.

Improvements in PCA tube quality, arc tube processing and mounting no longer require the use of this relatively cumbersome design.

The outer bulb is usually made of hard borosilicate glass, which is resistant to thermal shock. The bulb is then sealed to the mount assembly, evacuated, baked and lighted.

To remove the last vestiges of air left in the outer bulb after its

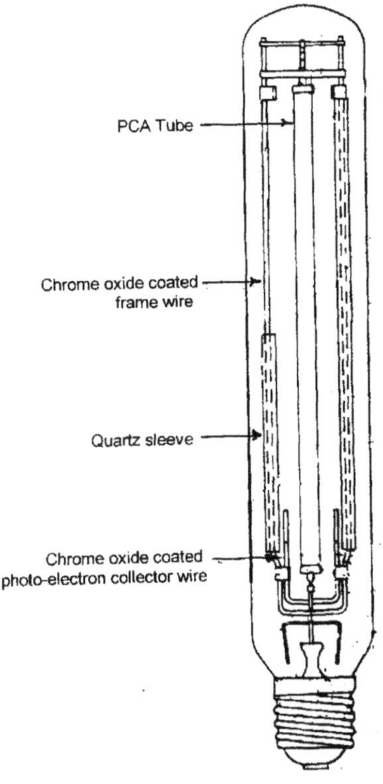

Figure 9-13. A mount construction designed to prevent the migration of sodium ions through the PCA wall as induced by the photo-electric effect (from Knochel et al, U.S. Patent 3,780,331, 1972).

evacuation, a barium getter is flashed onto the inside wall of the bulb at a location that is remote from the arc tube in order not to block the light. To do this, the getter rings are usually mounted close to the stem and the flashed getter is directed away from the arc tube's light-emitting surface. The flashing is done with a radio frequency induction coil. As the lamp ages, the activated getter will absorb the residual gases, primarily hydrogen, that slowly evolve from the arc tube and lamp components until it becomes saturated. A properly gettered lamp will show no color glow when it is excited with a high frequency induction coil.

The good vacuum produced by this process, minimizes the heat loss of the arc tube during lamp operation, thereby allowing it to operate at its maximum temperature.

Outer Bulb Shapes

The outer bulb is made in many sizes and shapes. There are three shapes that are most commonly used. One is elliptical, a shape often used when an inert fine powder, such as titania, silica or calcium pyrophosphate is deposited on the inner bulb wall. The powders themselves do not produce light, but are used to diffuse the light emitted from the arc tube so as to reduce any glare.

Tubular bulbs, which can be either of a single- or double-ended design, are generally used without a coating because the clear bulb and bright linear light source are well-suited for accurate optical control.

The third style is the reflector-shaped lamp. It is often made with an internal dichroic reflective coating which directs the light output out the front end, while letting the infrared energy escape through the back. The back end of the reflector bulb, emitting virtually no light, is usually mounted inside of a fixture.

Fusing High-Wattage Lamps

High-wattage lamps, such as the 700W and above, operate on ballasts having open circuit voltages in excess of 400V, and must be fitted with a fuse that will prevent a destructive arc from forming across the stem leads, if either the inner arc tube or outer bulb develops a leak. As soon as the air or arc tube vapors begin to fill the vacuum jacket, and a glow discharge is initiated, the metal shorting fuse will heat up and quickly melt in such a way so as to short-circuit the stem leads and render the lamp inoperative.[27] Another fuse that opens instead of shorts, can be added inside of the base to protect the bulb socket from damage in the unlikely event that the high open circuit voltage or the high starting pulse voltage cause an arc between the outer leads inside of the base chamber.[28]

Color and Output Characteristics

The high luminous efficacy of the sodium lamp is due largely to the location of its strong resonance lines lying close to the peak of the eye sensitivity curve, which is at 555 nm. At that wavelength, one watt of radiation would produce 683 lumens. At 589.0-589.6 nm, where the sodium lines appear, each watt of radiation yields about 500 lumens. This makes possible the high efficacy of the low pressure sodium lamp.

However, because the radiation is limited to such a narrow part of the visible spectrum, the color rendition is extremely poor.

Spectral Distribution

The HPS lamp, with its broader spectrum and high efficacy had, by 10 years after its introduction, become widely accepted for outdoor applications, but its low color rendition and color temperature left it undesirable for many other applications. The spectral power distribution of a typical 250W HPS lamp is shown in Figure 9-14.

It was known from the early work of Schmidt that increasing the sodium pressure would result in improved color rendition, but this resulted in reduced efficacy and life.

In 1978, an improved color 250W lamp was described by Bhalla, et al.[29] The reversal width of the sodium resonance lines was increased to 26 nm (see Figure 9-15). The arc length was shortened and the amalgam ratio was increased to 25% sodium.

This lamp had a luminous efficacy of 100 LPW and a CRI rating of 65, but the color temperature was still only 2100K. This lamp was satisfactory for a good number of indoor applications, but a higher color temperature was still needed to gain full market acceptance.

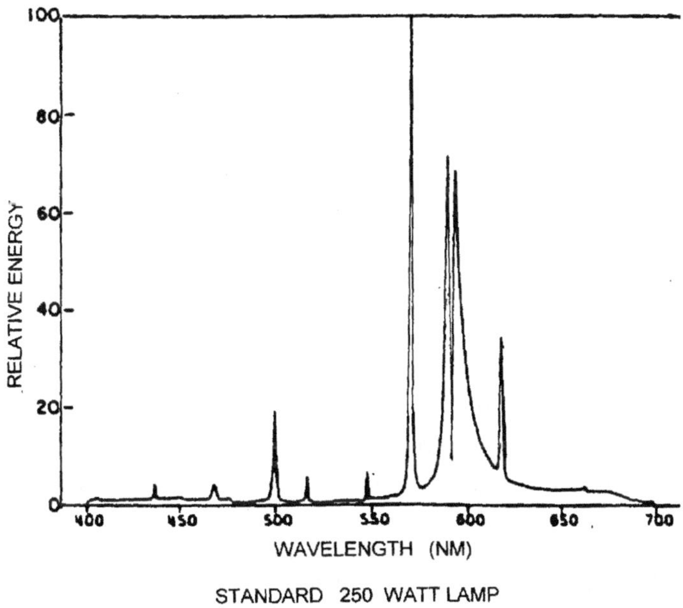

Figure 9-14. The spectral distribution of a typical 250W HPS lamp (from Bhalla et al, 1979). Courtesy of the Illuminating Engineering Society of North America.

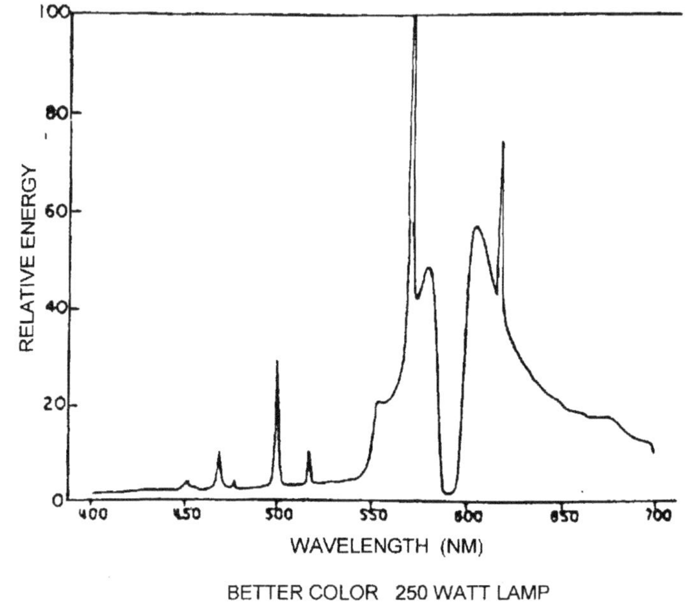

Figure 9-15. The change in the spectral distribution of a 250W lamp that results when the sodium vapor pressure and the sodium amalgam ratio are increased (from Bhalla et al, 1979). Courtesy of the Illuminating Engineering Society.

Super Color Lamps

As the sodium pressure increases, the color rendition continues to increase, until it reaches a maximum of 85 CRI at about 500 torr. Increasing the arc tube diameter also increases the spectral broadening and reversal widths. This produces a higher color rendition at lower sodium pressures.

Akutsu et al, in their 1984 paper, describe two groups of lamps which operate at 150, 250 and 400 watts. One group has a color rendition of 85 at a color temperature of 2500K, and for the 400W lamp, a luminous efficacy of 57 LPW.[30]

The other group, operating at a higher fill pressure, has a CRI rating of 78 at a color temperature of 2800K. The 400W lamp efficacy is 45 LPW.

Both groups were made with larger arc tube diameters, from 8mm for the 150W size to 13mm for the 400W size. The larger diameters also lower the temperature of the amalgam required at the cold spot to obtain the desired CRI rating. This is possible because a lower sodium pressure will produce the increased reversal width and broadening.

The authors also showed that increasing the arc tube wall loading from 20 to 41W per square centimeter increased the color temperature from 2556K to 2780K.

However, increasing the wall loading above 20W per square centimeter is not very practical, because chemical reactions at the wall will produce excessive sodium loss.

There are two ways that the wall loading can be increased while the temperature is kept low enough to limit the sodium reaction.

The first is to have a short spacing between the electrodes, so that the center of the arc tube is cooled down by conduction to the ends of the arc tube. This is shown by Ogato and Saito, who made a 50W lamp with a color temperature of 2500K, a CRI above 80 and a luminous efficacy of 40 LPW.[31] In this lamp, the electrode spacing is less than 10mm and the PCA tube is tapered at the ends.

The second method of lowering the wall temperature was done by Carlton et al, who placed a nitrogen gas into the outer bulb.[32] The 50W and 100W lamps they made have a color temperature of 2700K, a CRI rating of 85 and a luminous efficacy of 43 LPW.

The wall loading of these lamps is very high; 91W per square centimeter for the 100W lamp. This loading is made possible by the short electrode spacing and 500 mbar of nitrogen in the outer bulb. The "super-color" arc tube used was of the high pressure, shorter arc length design developed by Philips Lighting earlier.[33]

Normally, nitrogen cannot be used in the outer jacket since it reacts with niobium at the temperatures found in standard HPS arc tubes. However, in this arc tube, the niobium terminal lead is protected by a sleeve of PCA over most of its length. The small amount of niobium wire that is left exposed to the nitrogen has been lowered to a temperature under 500°C, where the reaction is not significant (see Figure 9-16).

Lamp Operation

The HPS lamp contains sodium and mercury. The mercury is added initially as a means of increasing the lamp voltage and it does this very well. A lamp having an amalgam which is 21% sodium by weight,

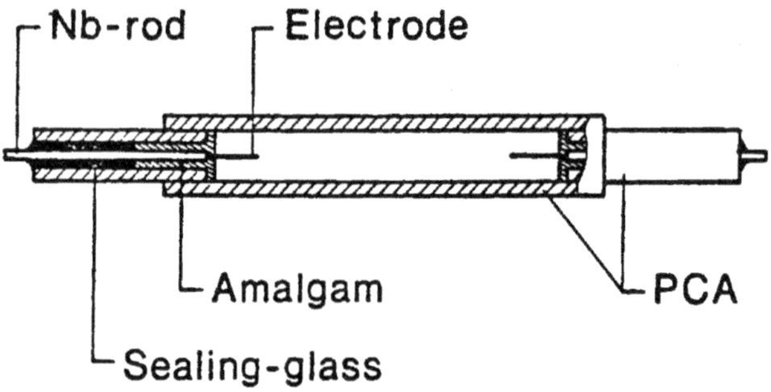

Figure 9-16. A compact, highly loaded, improved-color HPS arc tube (from van Kemenade et al, 1987). Courtesy of the Illuminating Engineering Society of North America.

will have about twice the voltage of a lamp having 100% sodium, if the sodium pressure is the same.

As stated earlier, the addition of mercury affects the long wavelength (red) side of the sodium resonance line by producing considerable spectral broadening. The atomic spectrum of mercury does not usually appear because it takes a higher temperature to excite the atoms. However, the yellow and green lines of mercury do appear when the sodium-to-mercury ratio is below 12%. When these lines appear in lamps dosed with a higher percentage of sodium, it indicates that some of the sodium has already been lost.

The amalgam is added in excess of the amount needed to provide the vapor pressure required for lamp operation. This is done so that the sodium that is lost to processing and operation will not affect the voltage or color too quickly.

As the percentage of sodium in the amalgam decreases, the voltage increases, if the sodium pressure remains the same. The loss of sodium during life results in increasing lamp voltage, and is one of the factors affecting lamp life.

The "Cold" Spot

The operation of the lamp is strongly affected by the temperature of the cold spot, the point in the arc tube at which the amalgam condenses, as shown in the curves of Figure 9-9. There we see that if the amalgam temperature changes from 660°C to 700°C, then the lamp voltage will change from 85V to 115V.

For many years, the cold spot was located at the end of the niobium exhaust tube which was most distant from the hot center of the arc tube and the glowing electrode tip. The heat flow from both of these thermal sources helped to maintain the cold spot temperature, although some of it was lost by radiation and conduction.

In many of the newer lamp designs there is no exhaust tube, so the cold spot lies within the arc tube behind the electrode.

In the lamps with nitrogen in the outer jacket there is cooling due to conduction and convection through the gas. As many as 20 factors have been identified that will affect the temperature of the cold spot.

The Ignitor

HPS lamps, because of their narrow tube diameter and xenon gas fill, require some assistance in starting that is well above the nominal open circuit voltage of the ballast. The starting is achieved with either an external or a ballast-enclosed ignitor.

However, some newer lamp designs have the ignitor built into the lamp itself. Like the other ignitors, it is a small electronic circuit that provides a starting pulse of 2500-4000V of about 1 microsecond duration. This pulse is superimposed on the 60 Hz open circuit voltage waveform. Its function is to ionize the xenon gas thereby allowing the lamp to start. As soon as starting occurs, the pulse is shut off by its own switching circuit.

Frit Seal Integrity

A prerequisite for reliable operation of PCA lamps is the overall quality of the arc tubes, but perhaps most of all, the frit seals. Each time the lamp is turned on or off, the arc tube is subjected to an enormous change in temperature, rising in a matter of minutes, from ambient temperature to 700°C at the end seals.

In addition, it must operate with a great temperature gradient along the tube wall, from 1200°C at the center to the 700°C at the ends. If the thermally induced stresses exceed the fracture stress of the seals, microcracks develop throughout the seal area, through which sodium will escape and condense on the outer envelope. This loss of sodium within the arc tube allows the mercury pressure to rise, which is another contributing cause to the gradual increase in lamp voltage.

Furthermore, the electrode will gradually deteriorate, its emission material will become depleted and it will begin to consume more power. This will increase the amalgam temperature and raise the voltage.

Lamp Cycling

Until recently, the voltage within the lamp did increase substantially as the lamp accumulated burning hours. Eventually, it would exceed the ability of the ballast to sustain it and the lamp would go out. After it had cooled down sufficiently to allow the vapor pressure to drop, arc reignition occurred. This phenomenon is referred to as "lamp-cycling," and whenever it happened, the lamp had to be replaced.

The new unsaturated lamps that have been developed use a very limited amount of sodium amalgam. The net result is that even though the proportion of mercury increases as the sodium is depleted, it is not enough to extinguish the lamp.[34,35]

Applications

HPS lamps are most generally used in outdoor lighting applications. These include roadway lighting, security lighting, walkways, and floodlighting of buildings and bridges. They have also found wide acceptance in many indoor industrial locations, and even some commercial areas, such as tennis courts and playgrounds. Their high lumen efficacy, long life and acceptable color make them suitable for places where economy in operation means more than exceptional color rendition. Their practically non-existent UV output allows the use of comparatively less expensive plastic refractors in luminaires.

Because of their relatively high intensity, small diameter and length, HPS lamps lend themselves to precise optical control, thereby enabling the use of higher mounting heights on roadways by reducing the amount of "spill" or unutilized light, and reducing the number of mounting poles required. Many highway interchanges, on-off ramps and similar roadway locations, where high mounting heights are preferred, have switched from fluorescent and low pressure sodium lamps to high pressure sodium lamps because they offer better brightness control, lower maintenance costs, improved visual comfort and lower installation costs.[36]

Many cities have converted to HPS lamps for their street lighting to take advantage of the reduced energy costs, and as an added bonus they have found that the higher light level light has helped to reduce crime[37] and increase public safety.[38]

For certain applications, where more diffuse light is desired, such as in open vertical fixtures, and where the specular glare of the light source is objectionable, a diffusing coating is applied to the inside surface of the outer bulb.

Kane and King developed lamps containing two arc tubes mounted in parallel (see Figure 9-17) to provide immediate re-lighting by the second (unlit) arc tube in the event that first lighted arc tube extinguishes, in case a momentary power interruption occurs.[39]

A 70W HPS lamp, mounted in a hard glass PAR38 outer bulb, was developed at the request of mine operators for use on underground equipment, but has also found use in many other applications. The lamp is made with a dichroic reflector that directs visible light out of the front of the bulb, but allows most of the heat to pass out of the rear. It is one of the first commercial high pressure lamps to be made with gas in the outer jacket, in this case, the gas is krypton.[40]

Studies have been made regarding the possible use of HPS lamps for vehicle headlamps.[41] The lamps must operate at no more than 30 watts of power, fit inside of existing sealed beam envelopes, and provide a beam of light comparable to the incandescent lamps currently in use. Unfortunately, problems of prolonged arc stabilization and restrike times still exist, and the added cost of lamp ballasting must be considered.

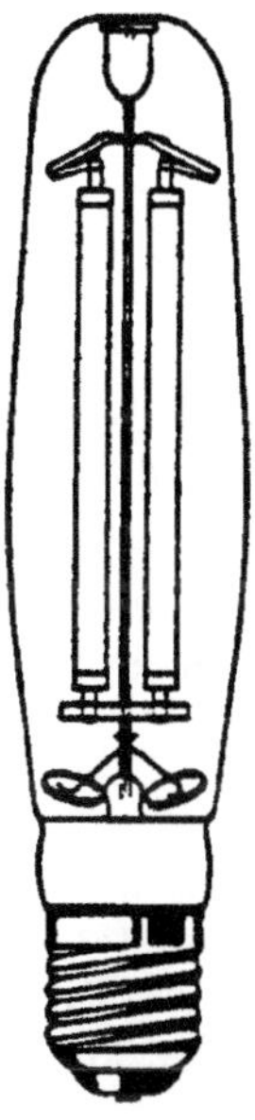

Figure 9-17. A 400W HPS lamp containing two arc tubes mounted in parallel (from Kane and King, 1986). Courtesy of the Illuminating Engineering Society of North America.

Looking Ahead

Much attention is now being given to the development of high pressure sodium lamps which will provide much better color rendition and encourage their use indoors.[42] There is already a substantial selection of improved color lamps available in lower wattages and smaller sizes.

Such lamps are now being made with a CRI rating of 85 and a color temperature of 2700K. Other progress in color is being made through the introduction of metal halide compounds into the arc stream, as is used so successfully in metal halide lamps.

With the improved color, the relatively small size of the arc tube encourages the development of narrow-beamed spotlights for innumerable, commercial high-lighting applications now using less efficient and shorter-lived incandescent lamps.

Extensive modifications are still being made to the design of the HPS lamp by numerous lamp development laboratories, to further improve this fine light source. Although HPS lamps are widely accepted for outdoor use, warehouses and industrial plants, the market is there to encourage their use in commercial and residential applications, because of their desirable characteristics of long life, good lumen maintenance and high light output.

Much effort is now being directed toward achieving greater indoor acceptance, by making further improvements in color rendition and correlated color temperature. To that extent, new designs such as the Philips "White SON" lamps and "Comfort Lamps" have already reached the marketplace.

As the cost of the HPS lamps and ballasts continue their downward trend, and as the color rendition improves, there is no question that there are still new horizons to be conquered by this ubiquitous light source!

References

1 Coble, R.L., "Transparent Alumina And Method Of Preparation." U.S. Patent 3,026,210, 1962.

2 Schmidt, K., "Metal Vapor Lamps." U.S. Patent 2,971,110, 1961.

3 Schmidt, K., "High Pressure Sodium Vapor Lamp." U.S. Patent 3,248,590, 1966.

4 Schmidt, K., "High Pressure Saturated Vapor Sodium Lamp Containing Mercury." U.S. Patent 3,384,798, 1968.

5 Louden, W.C. and Schmidt, K., "High Pressure Sodium Discharge Lamps," *J. Illum. Eng. Soc.*, July 1965.

6 Jack, A.G. and Koedam, M., "Energy Balances For Some High Pressure Gas Discharge Lamps," *J. Illum. Eng. Soc.*, July 1974.

7 Plumb, J., Duggan, G. et al, "Emitter For Unsaturated Vapor High Pressure Sodium Lamps," *J. Illum. Eng. Soc.*, 1995.
8 Peelen, J.G.J., "Light Transmission Of Sintered Alumina," *Philips Tech. Review*, 1976.
9 Louden, W.C. and Matz, W.C., "High Intensity Sodium Lamp Design Data For Various Sizes," *J. Illum. Eng. Soc.*, September 1966.
10 Luthra, K.L., "Sodium Loss Reactions In High Pressure Sodium Lamps," *J. Electrochem. Soc.*, January 1987.
11 Chu, G.P.K., "Properties And Evaluation Of HPS Lamp Seals: Materials And Structures," *J. Illum. Eng. Soc.*, July 1979.
12 Lehman, R.L. and Gurney, A., "Characterization Of HPS Arc Tube Seals," Dept. Of Ceramic Engineering, Rutgers University, 1983.
13 Waymouth, J.F. *Electric Discharge Lamps*. Cambridge, MA: MIT Press, 1971.
14 De Groot, J.J. and Van Vliet, J.A.J.M., *The High Pressure Sodium Lamp*. Philips Tech. Library, 1986.
15 Bhalla, R.S., "Improved Electron Emission Materials For HID Lamps," *J. Illum. Eng. Soc.*, April 1979.
16 Otani, K., Kawahara, K., Watanabe, K. and Tsuchihashi, M., "The Effect Of The Xenon Gas On The High Pressure Sodium Arc," *J. Light & Vision Env.*, Vol. 5-2: P. 1.
17 Plumb, J., Duggan, G. et al, Ibid.:P.26.
18 Jacobs, C.A.J., Sprengers, I. and De Vaan, R.L.C., "Arc Voltage Control In Low And High Pressure Sodium Lamps," *J. Illum. Eng. Soc.*, July 1978.
19 Louden, W.C., And Schmidt, K., "Intermediate Pressure Wall Stabilized Gas Lamp," U.S. Patent 3,054,992, 1962.
20 Louden, W.C. and Pinter, R.S., "Ceramic Lamp Construction," U.S. Patent 3,243,635, 1966.
21 Beijer, L.B., Van Boort, H.J.J., Koedam, M., "The Sodium Discharge Lamp," *Lighting Design & Application*, July 1974.
22 Schmidt, K., Ibid., U.S. Patent 3,384,798, 1968.
23 Denbigh, P.L., "Experimental Approach To HPS Lamp Design," *Lighting Research & Technology*, V.10-1(1978):P.31.
24 Jacobs, C.A.J., Sprengers, I., and De Vaan, R.L.C., Ibid:P.127.
25 Cohen, S. and Richardson, D.A., "The Application Of Penning Mixture Starting Gases To High Pressure Sodium Lamps," *Lighting Design & Appl.*, September 1975.
26 Knochel, W.J., Werner, L.C. and Lin, F.C.M., "Apparatus & Method

For Eliminating Microcracks In Alumina Discharge Devices, Etc.," U.S. Patent 3,780,331, 1972.
27 Collins, B.R., Sules, J. and Mcvey, C.I., "High Intensity Lamp Containing Internal Shorting Fuse," U.S. Patent 3,767,965, 1973.
28 Collins, B.R., "High Intensity Lamp Containing Arc Extinguishing Base," U.S. Patent 3,849,691, 1974.
29 Bhalla, R.S., Larson, D.A. and Unglert, M.C., "HPS Lamp With Improved Color Rendering," *J. Illum. Eng. Soc.*, July 1979.
30 Akutsu, H., Watarai, Y., Saito, N., and Mizano, H., "A High Pressure Sodium Lamp With High Color Acceptability," *J. Illum. Eng. Soc.*, July 1984.
31 Ogata, Y. And Saito, N., "A New 50 Watt Super High Pressure Sodium Lamp," *J. Illum. Eng. Soc.*, 1988.
32 Carlton, S., Keijser, R.A.J., Van Kemenade, J.T.C., "White HPS Lamps With A Color Temperature Of 2700K," *J. Illum. Eng. Soc.*, 1991.
33 Van Kemenade, J.T.C., Keijser, R.A.J. and Bolt, B.F., "New Possibilities For HPS Lamps In Indoor Lighting," *J. Illum. Eng. Soc.*, 1987.
34 Ingalls, P., Dolan, R., and Van Bockstal, G., "Characteristics Of Unsaturated HPS Lamps Operated On Mercury Ballasts," *J. Illum. Eng. Soc.*, 1993.
35 Ravi, J., Luijks, G., "A Non-Cycling High Pressure Sodium Lamp," *J. Illum. Eng. Soc., 1995.*
36 Furst, W.L., "Sodium Discharge Lamps And Their Application In Roadway Lighting," *Canadian General Electric Bulletin*: HID-Lij, February 1970.
37 Le Vere, C.W., "Street Lighting As A Crime Deterrent," *Metropolitan State College Report*, November 1976.
38 Practical Methods, "How Baltimore Is Lighting Its Streets," *Electrical Construction & Maint.*, November 1975.
39 Kane, R.M. and King, N.R., "A 400 Watt Instant Restrike Double Arc Tube HPS Lamp," *Lighting Design & Appl.*, December 1986.
40 Gibson, R.G., "Design Features Of A 70 Watt Hps Gas-Filled Reflector Lamp," *J. Illum. Eng. Soc.*, 1991.
41 Szekacs, G., "Low Wattage Hps Lamps For Vehicle Headlamps," *J. Illum. Eng. Soc.*, 1986.
42 Gorzelnik, E.F., "HPS Units Save Energy In Offices, Plants," *Electrical World*, June 15, 1978.

Index

For Product Safety Concerns and Information please contact our EU representative GPSR@taylorandfrancis.com Taylor & Francis Verlag GmbH, Kaufingerstraße 24, 80331 München, Germany

Batch number: 10725935

Printed by Printforce, the Netherlands